T0192307

SpringerBriefs in Electrical and Computer Engineering

More information about this series at http://www.springer.com/series/10059

Cam Nguyen · Joongsuk Park

Stepped-Frequency Radar Sensors

Theory, Analysis and Design

Springer

Cam Nguyen
Texas A&M University
College Station, TX
USA

Joongsuk Park
Digital Media R&D Center
Samsung Electronics Co., Ltd.
Suwon
Korea, Republic of (South Korea)

ISSN 2191-8112 ISSN 2191-8120 (electronic)
SpringerBriefs in Electrical and Computer Engineering
ISBN 978-3-319-12270-0 ISBN 978-3-319-12271-7 (eBook)
DOI 10.1007/978-3-319-12271-7

Library of Congress Control Number: 2016935214

Printed on acid-free paper

This Springer imprint is published by Springer Nature
The registered company is Springer International Publishing AG Switzerland

Preface

Stepped-frequency radar sensors are attractive for various surface and subsurface sensing applications. Stepped-frequency systems transmit consecutive trains of continuous-wave signals at different frequencies that are separated by a fixed amount. The unique characteristic of stepped-frequency systems is, although they work as frequency-based systems, their final response is described in a time-domain quantity, namely "synthetic pulse," which contains the information about targets. This unique "synthetic pulse" product enables stepped-frequency radar sensors to have some of the advantages of impulse-based ultra-wideband (UWB) systems, which operate completely in the time domain, such as ease in identification and characterization of adjacent targets. Specifically, stepped-frequency radar sensors have several major advantages. First, they have a very narrow instantaneous bandwidth at each frequency, resulting in less receiver's noise figure and hence increased sensitivity and dynamic range. Second, their absolute RF operating bandwidth, on the other hand, can be very wide, leading to fine range resolution. Third, they can transmit high average power enabling long range or deep penetration. Fourth, it is possible to properly shape the transmitted spectrum by transmitting constituent signals with certain amplitudes and phases, helping improve the system's performance and possibly compensate for inevitable effects due to system's imperfection and operating environment. Stepped-frequency radar sensors find numerous applications for military, security, civilian, commerce, medicine, and healthcare.

This book presents the theory, analysis, and design of stepped-frequency radar sensors and their components. Specifically, it addresses the following main topics of stepped-frequency radar sensors: system analysis, transmitter design, receiver design, antenna design, and system integration and test. It also presents the development of two practical stepped-frequency radar sensors and their transmitters, receivers, antennas, signal processing, integrations, electrical tests, and sensing measurements, which serve as an effective way to demonstrate not only the analysis, design, test, and sensing applications of stepped-frequency radar sensors, but also the design of constituent components. Although the book is succinct, the material is very much self-contained and contains practical, valuable, and sufficient information presented in such a way that allows readers with an undergraduate background in electrical engineering or physics, with some experiences or graduate courses in microwave circuits, to understand and design easily stepped-frequency radar sensors and their transmitters, receivers, and antennas for various sensing applications.

The book is useful for engineers, physicists, and graduate students who work in radar, sensor, and communication systems as well as those involved in the design of RF circuits and systems. It is our hope that the book can serve not only as a reference for the development of stepped-frequency systems and components, but also for possible generation of ideas that can benefit many existing sensing applications or be implemented for other new applications.

College Station, TX; Newport Beach, CA, USA — Cam Nguyen
Seoul, South Korea — Joongsuk Park

Contents

Chapter 1
Introduction

Radar sensors, in general, are a very mature topic with many developments since before World War II. The word RADAR stands for RAdio Detection And Ranging and was first used by the United States Navy in 1940. The concept of radar, however, could be considered as taking place around 1886 when Heinrich Hertz demonstrated the transmission, reception, and reflection of electromagnetic waves. Radar sensors have advanced significantly from their beginning as detection devices for military applications in World War II and have extended their usage for many non-military applications across radio-frequency (RF) spectrums from low to microwave, millimeter-wave, and submillimeter-wave frequencies, such as sensing abnormal human body condition, diagnosing it and imaging the effect; early detection of cancer; detection and inspection of buried underground objects; and detection, tracking and monitoring of hidden activities, etc.

Some of the most significant usages of radar sensors for non-military applications are for surface and subsurface sensing. In these areas of sensing, radar sensors have been used as a fast, reliable, accurate and cost-effective technique for non-destructive, non-contact sensing of surfaces and subsurfaces in various applications such as detecting and locating buried objects, tunnels, archeological sites; measuring distances, displacements, material's thickness and moisture content; evaluation of wood composites; detection and inspection of underground pipe infrastructures; and inspection of bridges, pavements and other civil infrastructures, e.g., [1–12]. The radar sensors for surface and subsurface sensing typically implement impulse technique and continuous-wave (CW) methods with different frequency modulations for transmitting signals like frequency-modulated continuous wave (FMCW) or stepped-frequency continuous wave (SFCW).

The most distinguished difference between an impulse system and a FMCW or SFCW system (and in general other non-impulse or CW based systems) is in the transmitting waveform. An FMCW system transmits and receives CW (sinusoidal) signals, one signal at each frequency, subsequently across a bandwidth. An FMCW system does not transmit and receive signals of different frequencies simultaneously.

C. Nguyen and J. Park, *Stepped-Frequency Radar Sensors*,
SpringerBriefs in Electrical and Computer Engineering,
DOI 10.1007/978-3-319-12271-7_1

That is, an FMCW system is basically operated over a bandwidth of single-frequency signals. A SFCW system transmits consecutive trains of CW signals, each at different frequency separated by a certain amount, toward targets and receives reflected signals from the targets. The received digital in-phase and quadrature signals from a target are then transformed into a synthetic pulse in time domain using inverse discrete Fourier Transform, which contains the target's information. A SFCW system has very narrow instantaneous bandwidth at each frequency, resulting in desirably high signal-to-noise ratio at the receiver. Its entire bandwidth, on the other hand, can be very wide, leading to fine resolution. Moreover, its high average transmitting power enables long range or deep penetration. Although the final received signal of a SFCW system is transformed into a time-domain pulse signal, it is still CW-based and does not transmit and receive signals of different frequencies simultaneously. On the other hand, an impulse system transmits and receives a periodic (non-sinusoidal) impulse-type signal, which contains many constituent signals occurring simultaneously with each having a different frequency. In other words, an impulse system transmits and receives many CW signals having different frequencies concurrently. It is this characteristic that makes the impulse system and FMCW or SFCW system (and other CW based systems) distinctively different in the system architecture, design, operation, performance, and possible applications.

An overview of the fundamentals of the impulse, FMCW and SFCW radar sensors is provided as follows.

1.1 Impulse Radar Sensors

The impulse radar sensors, also known as the pulsed or time-domain radar sensors, typically employ a train of pulses (loosely considered as impulses), mono-pulses (or mono-cycle pulses), or modulated pulses as the transmitting waveform. Figure 1.1 describes some of these waveforms. One of the first radar sensors for subsurface sensing is the impulse radar that measured the properties of coal [13]. As illustrated in Fig. 1.1, an impulse radar sensor transmits a short pulse train with a pulse repetition interval (*PRI*). Such an impulse can be generated directly by using transistors, step recovery diodes (SRD), or tunnel diodes, e.g., [14–18]. It is noted, according to the Fourier series, that a periodic impulse can also be generated indirectly from individual concurrent sinusoidal signals having appropriate amplitudes and phases at different frequencies, and such an impulse system can be considered as equivalent to a frequency-domain system. However, this technique is very difficult to be implemented in practice at microwave frequencies due to the difficulty in generating individual sinusoidal waveforms at different frequencies with precise amplitudes and phases.

Impulse radar sensors have been widely used in many sensing applications owing to their simple architecture and design. The range (or vertical) resolution ΔR of a system, which indicates how close targets within a specific range R can be distinguished, is given by

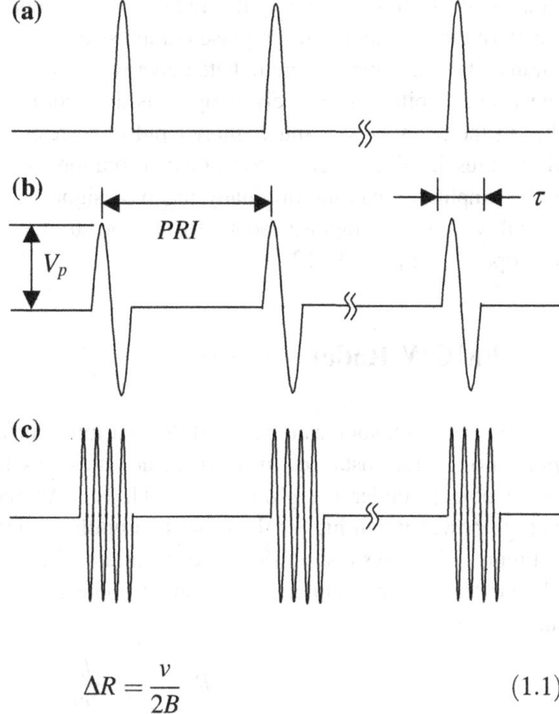

Fig. 1.1 Typical waveforms of impulse radar sensors: **a** impulse, **b** mono-pulse (τ is the pulse width and V_p is the peak amplitude), and **c** modulated pulse

$$\Delta R = \frac{v}{2B} \quad (1.1)$$

where $v = c/\sqrt{\varepsilon_r}$ with $c = 3 \times 10^8$ m/s and ε_r being the speed of light in air and relative dielectric constant of the propagating medium, respectively, is the velocity of the signal and B is the absolute operating bandwidth of the system, which is essentially the absolute RF bandwidth of the transmitted signal. The range resolution is inversely proportional to the bandwidth. Impulse radar sensors typically have much wider bandwidths than those of CW based systems due to the extremely wideband nature of the impulse-type signals, facilitating high resolution. These signals contain both low and high frequency components, making long range possible due to small attenuation at low frequencies. Impulse radar sensors are hence suitable for applications requiring fine range-resolution and/or long range, provided that they can be designed with sufficiently short duration and/or high peak voltage. Ultra-short or high-power pulses are restricted by available device technologies and are difficult to design, particularly when both ultra-short pulse and high peak-voltage are needed simultaneously. Consequently, the use of impulse radar sensors for very high resolution and/or long range applications could be limited in practice. It should be noted that this problem does not exist only for impulse radar sensors; it is also difficult to design extremely wide-band and high-power CW systems for very high resolutions and long ranges. Moreover, the inherently large RF bandwidths of narrow pulses degrade the receiver's noise figure of impulse radar sensors, which results in decreased sensitivity and hence dynamic range for the receiver. It should also be noted that simultaneous achievement of both long range and high range resolution can be facilitated by employing the pulse compression technique for impulse radar sensors. This, however, increases the complexity of the system design. In pulse-compression impulse radar sensors, a long pulse is transmitted to achieve large radiated energy needed for long range and hence better detection capability. The received signal is then compressed by a pulse-compression filter in the receiver to a much shorter pulse, corresponding to a much wider bandwidth, thus leading to enhanced range resolution. The compressed pulse also has higher amplitude than the originally received signal, further improving the detection capability. Various impulse radar sensors for surface and subsurface sensing were developed, e.g., [1–3, 9, 12].

1.2 FMCW Radar Sensors

FMCW radar sensors can be used for various surface and subsurface sensing applications—for instance, measuring the thickness of coal layers and detecting buried objects under the ground [19–21]. FMCW radar sensors can achieve an average power much higher than that of impulse radar sensors.

Figure 1.2 shows a simplified bock diagram of FMCW radar sensors employing a linear frequency-modulation transmitter. The range R of a target can be determined from

$$R = \frac{v\tau}{2} = \frac{vf_d}{2m} \tag{1.2}$$

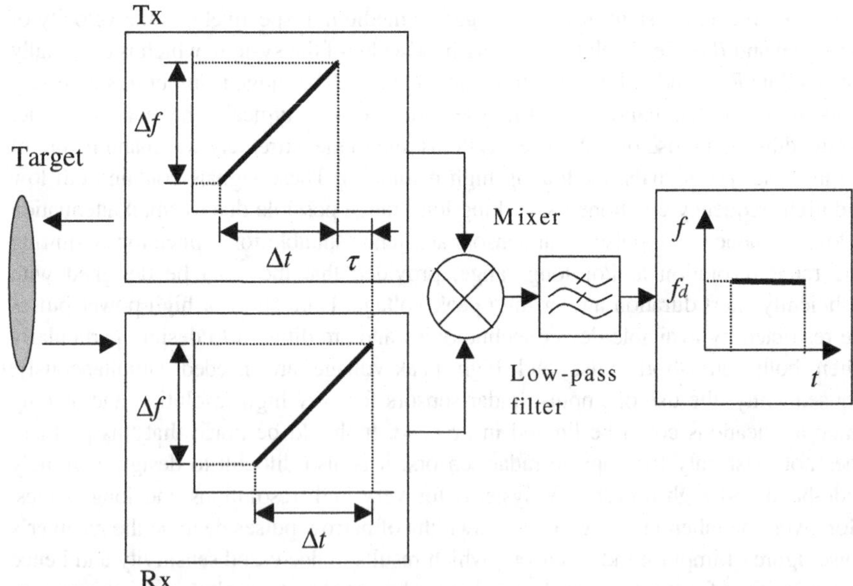

Fig. 1.2 A simplified FMCW radar sensor illustrating the linear-modulation transmitted and received frequencies, and the beat frequency

where f_d is the beat frequency resulting from the relative time delay τ between the transmitted and returned signals and down-converted by the mixer, and m is the rate of sweeping frequency.

For a given range and propagating medium, the accuracy of the range measurement depends on that of the frequency's sweeping rate, which is an important design parameter for FMCW radar sensors. In practice, it is difficult to achieve an accurate and constant frequency's sweeping rate over a wide band due to the non-linearity of the synthesizer, particularly when a voltage-controlled oscillator (VCO) is used in its place. Moreover, a wide bandwidth degrades the receiver's noise figure, which results in decreased sensitivity and hence dynamic range for the receiver. These drawbacks may prohibit possible use of FMCW radar sensors for some applications that require wideband operation with very high range accuracy.

1.3 SFCW Radar Sensors

The concept of the stepped-frequency technique was first introduced in 1972 by Robinson et al., in which a SFCW radar sensor was realized and used to detect buried objects [22]. Active research in SFCW radar sensors, however, only began in the early 1990s.

SFCW radar sensors transform the amplitudes (A_i) and phases (ϕ_i) of the base-band in-phase (I) and quadrature (Q) signals in frequency domain to a synthetic pulse in time domain to find the range R of a target [23] as defined by

$$I_i = A_i \cos\phi_i = A_i \cos\left(-\frac{2\omega_i R}{v}\right) \tag{1.3}$$

and

$$Q_i = A_i \sin\phi_i = A_i \sin\left(-\frac{2\omega_i R}{v}\right) \tag{1.4}$$

where ω_i is the frequency. SFCW radar sensors have several advantages [24, 25]. First, they have a narrow instantaneous bandwidth that significantly improves the receiver's noise figure, and hence sensitivity and dynamic range, while maintaining good average power. Second, it can transmit a high average power, resulting in a long range or deep penetration, due to the use of CW signals. Third, the non-linear effects caused by the inherent imperfections of the transmitter and receiver can be conveniently corrected through appropriate signal processing. Fourth, as the system transmits only one frequency at a particular time, it facilitates accurate compensation for the received signals propagating through dispersive and lossy media (with known properties) through signal processing. Fifth, the system's Analog-to-Digital (A/D) converter uses a very low sampling frequency due to low frequency of the base-band I/Q signals, enabling great precision and ease in circuit design. Lastly, the transmitted spectrum can be properly shaped by transmitting signals with certain amplitudes and phases, which help improve further the system's performance such as resolution and signal-to-noise ratio, which in turn helps compensating for the constituting component's deficient responses at some frequencies as well as other unavoidable effects from the system and operating environment such as losses in propagating media. However, it should be noted that, it is very difficult in practice to design a microwave transmitter that can transmit signals with specified amplitudes and phases at different frequencies, especially at millimeter-wave frequencies. SFCW radar sensors, however, are complex and hence difficult to design and costly. For many sensing applications, these disadvantages are outweighed by the advantages mentioned earlier, making SFCW radar sensors an attractive solution for these applications, which in turn generates significant research and use of SFCW radar sensors.

Various SFCW radar sensors have been developed for sensing applications. Several of them were designed specifically for surface and subsurface sensing, e.g., [3, 8, 10, 11, 26–29]. The SFCW radar sensor operating in the 1–2 GHz range in [3] was developed for detecting landmines. The Ka-band (26.5–40 GHz) SFCW radar sensor in [8] was developed for distance measurement. The microwave SFCW radar sensor operating from 0.6 to 5.6 GHz was reported in [10] for characterization of pavement subsurface. Another millimeter-wave SFCW radar sensor working from 29.72 to 37.7 GHz was developed in [11] for surface and subsurface sensing. The SFCW radar sensor in [26] operated from 0.6 to 1.112-GHz and was developed for detecting moisture content in the pavement subgrade. The SFCW radar sensor in [27] was developed from 490 to 780 MHz for detection of buried objects. The SFCW system in [28] worked from 10 to 620 MHz. The SFCW radar sensor reported in [29] was configured for detecting concrete cracks using a commercially available network analyzer and operated from 0.5 to 6 GHz.

SFCW radar sensors find numerous applications for military, security, civilian, commerce, and medical and health care. The following describe some of these possible applications:

Military and Security Applications: detection, location and identification of targets such as aircrafts, tunnels, concealed weapons, hidden illegal drugs, buried mine and unexploded ordnance (UXO); locating and tracking personnel; detection and identification of hidden activities; through-wall imaging and surveillance; building surveillance and monitoring.
Civilian and Commercial Applications: detection, identification and assessment of abnormal conditions of civil structures such as pavements, bridges, buildings, buried underground pipes; detection, location and identification of objects;

measurement of liquid volumes and levels; inspection, evaluation and process control of materials; geophysical prospecting, altimetry; collision and obstacle avoidance for automobile and aviation.
Medical and Health Care Applications: detection and imaging of tumors; health examination of patients; medical imaging.

This book covers SFCW radar sensors. There are many aspects in SFCW radar sensors and a complete coverage would require a book of substantial size. The objective of this book is not to provide a full coverage of SFCW radar sensors. It is indeed not possible to have a complete discussion of SFCW radar sensors in this short book. It is our intention to address only the essential parts of SFCW radar sensors including system and component analysis, design, signal processing, and measurement in a concise manner, yet with sufficient details, to allow the readers to understand and design SFCW radar sensors for their intended applications, whether for research or for commercial usage.

The book presents the theory, analysis and design of SFCW radar sensors and their components in a limited scope, yet with essential details. It addresses the following main topics of SFCW radar sensors: system analysis, transmitter design, receiver design, antenna design, signal processing, and system integration and test. The developments of two SFCW radar sensors and their components using microwave integrated circuits (MICs) and microwave monolithic integrated circuits (MMICs), as well as signal processing, are included in sufficient details to demonstrate the theory, analysis, and design of SFCW radar sensors. Various measurements in surface and subsurface sensing for these systems are presented to verify the system workability. These measurements also serve as examples to demonstrate not only some specific usages of the sensors for surface and subsurface sensing, but also possible implementations for other sensing needs.

The book is organized as follows. Chapter 1 gives the introduction of SFCW radar sensors and some of their possible applications. Chapter 2 addresses the general analysis of radar sensors. Chapter 3 covers the analysis of SFCW radar sensors. Chapter 4 presents the development of a microwave and millimeter-wave SFCW radar sensors using MICs and MMICs including the designs of the transmitters, receivers and antennas, and signal processing. Chapter 5 discusses the electrical tests and sensing measurements of the developed microwave and millimeter-wave SFCW radar sensors. Finally, Chap. 6 gives the summary and conclusion.

Chapter 2
General Analysis of Radar Sensors

2.1 Introduction

The operation and performance of radar sensors involve the propagation of electromagnetic (EM) waves (or signals as commonly referred to). The transmitter of a radar sensor transmits a signal toward a target or object. The signal, upon incident on the target, scatters in all directions due change of the electric and/or magnetic properties of the target relative to those of the medium surrounding the target, in which the signal is propagating. The scattered signals propagating in the direction of the receiving antenna are captured and processed by the receiver. Understanding the behavior of signals and analysis of scattered signals are hence important in order to understand the operation and performance of radar sensors as well as characterize the properties of targets, which are needed in the design and operation of radar sensors.

Two of the most important characteristics dictating the performance of radar sensors are "resolution" and "range" (or "penetration depth" as typically used in subsurface sensing). The angle or cross or lateral resolution depends on the antenna while the range resolution is determined by the absolute bandwidth of the signal or, specifically for stepped-frequency continuous-wave (SFCW) radar sensors, the width of the synthetic pulse corresponding to a target that a SFCW) system generates. These are discussed in details in Chap. 3. On the other hand, the range is determined by various parameters and is discussed in this chapter. Some of these parameters are determined by the designer and used in the design of the system and its constituent components such as transmitter's power, antenna gain, signal's frequency, and receiver's gain, noise figure, dynamic range and sensitivity. Some of these parameters are related. Other parameters are dependent upon the properties of the media, in which the signal propagates, and individual targets. The properties of the media directly affect the propagation constant involving the loss and velocity of the propagating signal, which essentially dictate how signal travels in the media.

C. Nguyen and J. Park, *Stepped-Frequency Radar Sensors*,
SpringerBriefs in Electrical and Computer Engineering,
DOI 10.1007/978-3-319-12271-7_2

Targets have their own properties and Radar Cross Section (RCS), and cause reflection, transmission, spreading loss, and scattering of incident signals. These parameters are not controlled by the designer. However, if known, they could provide valuable information to the design of the radar sensor's architecture and components and hence are important to the design of radar sensors.

This chapter addresses various topics concerning the analysis of radar sensors including signal propagation, which involves Maxwell's and wave equations, propagation constant; signal scattering from objects, which involves reflection, transmission, radar cross section; system equations including Friis transmission equation and radar equations in general and for half-space and buried targets; signal-to-noise ratio; receiver sensitivity; maximum range or penetration depth; and system performance factor.

2.2 Signal Propagation

The propagation of signals in a medium is governed by Maxwell's equations, or the resulting wave equations, and the medium's properties. These will be briefly covered in this section assuming steady-state sinusoidal time-varying signals.

2.2.1 *Maxwell's Equations and Wave Equations*

Maxwell's equations in differential (or point) form are

$$\nabla \times \vec{E} = -j\omega\vec{B} \tag{2.1a}$$

$$\nabla \times \vec{H} = \vec{J} + j\omega\vec{D} \tag{2.1b}$$

$$\nabla . \vec{D} = \rho \tag{2.1c}$$

$$\nabla . \vec{B} = 0 \tag{2.1d}$$

where $\vec{E}$, $\vec{H}$, $\vec{D}$, $\vec{B}$, $\vec{J}$, and ρ are the (phasor) electric field, magnetic field, electric flux density (or displacement field), magnetic flux density, current density, and volume charge density. They are functions of frequency and location.

The electric flux density, electric field intensity and the magnetic flux density, magnetic field intensity in a material are related by the following constitutive relations:

$$\vec{D} = \varepsilon_o \varepsilon_r \vec{E} = \varepsilon \vec{E} \tag{2.2}$$

and

$$\vec{B} = \mu_o \mu_r \vec{H} = \mu \vec{H} \tag{2.3}$$

where $\varepsilon_o = 8.854 \times 10^{-12}\,\mathrm{F/m}$ and $\mu_o = 4\pi \times 10^{-7}\,\mathrm{H/m}$ are the dielectric constant or permittivity and permeability of air, ε_r and ε are the relative dielectric constant (or relative permittivity) and dielectric constant (or permittivity) of material, respectively, and μ_r and μ are the relative permittivity and permittivity of material, respectively. Most materials are non-magnetic having μ_r close to 1. Also, most materials are "simple" media, which are linear, homogeneous, and isotropic having constant ε_r and μ_r.

The (conduction) current is related to the electric field by

$$\vec{J} = \sigma \vec{E} \tag{2.4}$$

where σ is the conductivity of material.

Although the electric and magnetic fields of signals in any medium, and hence signal propagation, can be determined from Maxwell's equations subject to boundary conditions, it is more convenient to determine them from (single) wave equations. The wave equations can be derived from Maxwell's equations as

$$\nabla^2 E - \gamma^2 E - j\omega\mu J - \frac{1}{\varepsilon}\nabla\rho = 0 \tag{2.5a}$$

$$\nabla^2 H - \gamma^2 H - \nabla \times J = 0 \tag{2.5b}$$

where $\gamma = \alpha + j\beta$ is the propagation constant with α (Np/m) and β (rad/m) being the attenuation and phase constant, respectively.

2.2.2 Propagation Constant, Loss and Velocity

In general, a medium is characterized by its complex dielectric constant or complex permittivity, which, in turn, results in a complex propagation constant for signals propagating in the medium. The real and imaginary parts of the propagation constant are known as the attenuation and phase constants and dictate the loss and velocity of signals, respectively. Practical media in which signals propagate are always lossy and dispersive, and hence are imperfect. Consequently, there is always loss present in any practical medium, known as dielectric loss, due to a non-zero conductivity of the medium. This loss reduces the transmitting power and hence the maximum range or penetration depth of a radar sensor. The velocity, on the other hand, determines the target's range.

The dielectric properties of a medium, including (practical) lossy and (ideal) lossless media, can be described by a complex dielectric constant or complex permittivity of the medium ε_c given as

$$\varepsilon_c \equiv \varepsilon' - j\varepsilon'' = \varepsilon - j\frac{\sigma}{\omega} \tag{2.6}$$

where

$$\begin{aligned} \varepsilon' &= \varepsilon = \varepsilon_o \varepsilon_r \\ \varepsilon'' &= \frac{\sigma}{\omega} \end{aligned} \tag{2.7}$$

Both ε' and ε'' are functions of frequency, and ε'' accounts for the loss in the medium. We can also characterize a lossy medium by its complex relative dielectric constant

$$\varepsilon_{cr} \equiv \frac{\varepsilon_c}{\varepsilon_o} = \varepsilon_r' - j\varepsilon_r'' = \varepsilon_r - j\frac{\sigma}{\omega\varepsilon_o} \tag{2.8}$$

Note that $\varepsilon_r' = \varepsilon_r$.

The loss in a medium is typically characterized in term of the loss tangent defined as the ratio between the imaginary and real parts of the complex dielectric constant:

$$\tan\delta \equiv \frac{\varepsilon''}{\varepsilon'} = \frac{\varepsilon_r''}{\varepsilon_r'} = \frac{\sigma}{\omega\varepsilon} = \frac{\sigma}{\omega\varepsilon_o\varepsilon_r} \tag{2.9}$$

which is of course dependent upon frequency. The loss tangent of a medium, just like its relative dielectric constant ε_r and complex relative dielectric constant ε_{cr}, can be measured.

The propagation constant of signals travelling in a lossy medium can be derived as

$$\begin{aligned} \gamma &= \alpha + j\beta = j\omega\sqrt{\mu\varepsilon_c} \\ &= j\omega\sqrt{\mu\varepsilon}\sqrt{1 - j\frac{\sigma}{\omega\varepsilon}} = j\omega\sqrt{\mu\varepsilon'}\sqrt{1 - j\frac{\varepsilon''}{\varepsilon'}} = j\omega\sqrt{\mu\varepsilon}\sqrt{1 - j\tan\delta} \end{aligned} \tag{2.10}$$

or

$$\gamma = jk_o\sqrt{\varepsilon_r' - j\varepsilon_r''} = jk_o\sqrt{\varepsilon_r'}\sqrt{1 - j\frac{\varepsilon_r''}{\varepsilon_r'}} \tag{2.11}$$

where $k_o = \omega\sqrt{\mu_o\varepsilon_o}$ is the wave number for air, considering non-magnetic medium having $\mu_r = 1$.

The velocity of a signal can be determined from the phase constant as

$$v = \frac{\omega}{\beta} \tag{2.12}$$

For media having small loss, $\sigma/\omega\varepsilon \ll 1$ or $\varepsilon_r'' \ll \varepsilon_r'$ and hence $\tan\delta \ll 1$, and the propagation constant in (2.10) and (2.11) can be approximated using the binomial series as

$$\gamma = \alpha + j\beta = \frac{k_o\varepsilon_r''}{2\sqrt{\varepsilon_r'}} + jk_o\sqrt{\varepsilon_r'} = \frac{\omega\sqrt{\mu_o\varepsilon'}}{2}\tan\delta + j\omega\sqrt{\mu_o\varepsilon'} \tag{2.13}$$

and the velocity is obtained as

$$v = \frac{1}{\sqrt{\mu_o\varepsilon_o\varepsilon_r}} = \frac{c}{\sqrt{\varepsilon_r}} \tag{2.14}$$

where $c = 3 \times 10^8$ m/s is the velocity in air.

For high-loss media, $\sigma \gg \omega\varepsilon'$ or $\tan\delta \gg 1$, we can obtain from (2.10) and (2.11):

$$\gamma = \alpha + j\beta = j\omega\sqrt{\mu\varepsilon}\sqrt{j\frac{\sigma}{\omega\varepsilon}} = \sqrt{j}\sqrt{\omega\mu\sigma} = (1+j)\sqrt{\pi f\mu\sigma} \tag{2.15}$$

and the velocity is given by

$$v = 2\sqrt{\frac{\pi f}{\mu_o\sigma}} \tag{2.16}$$

The developed microwave and millimeter-wave SFCW radar sensors to be presented in Chaps. 4 and 5 were used for some surface and subsurface sensing applications. One of the subsurface sensing measurements was done on pavement structures. A typical pavement consists of three layers: asphalt, base, and subgrade. The subgrade is basically natural soil of infinitely thick. The common pavement materials used in practice could be considered as low-loss and non-magnetic materials [30, 31]. Table 2.1 shows typical parameters of the asphalt and base [9]. The values of the real (ε_r') and imaginary (ε_r'') parts of the complex relative dielectric constant were measured at 3 GHz using a vector network analyzer based dielectric measurement system. These measurements were done over many samples in a laboratory setting.

Table 2.1 Typical electrical properties of pavement materials

Parameter	Asphalt	Base
ε_r'	5–7	8–12
ε_r''	0.035	0.2–0.8

The values of the parameters in Table 2.1 show that these pavement materials can be considered as low-loss materials. It is noted that these measurements were done in laboratory. In field environments, where a radar sensor is used for sensing, however, the situation is typically more complex, resulting in different material properties. The properties of these materials, characterized by their complex relative dielectric constants, are influenced significantly by environmental factors such as rain, snow, moisture, and temperature, etc. They are also a function of frequency, space in the structures, and the constituents of materials. In addition, the property of a given material may also vary substantially from one sample to another. These factors make the actual complex relative dielectric constants different from what measured in a laboratory and the actual materials could be very lossy.

Practical media such as pavement materials are dispersive, causing the phase constant β behave as a nonlinear function of frequency, which in turn results in frequency-dependent velocities as can be inferred from (2.12). As such, signals of different frequencies in a radar sensor travel with different velocities. The return signals from a target hence have different phases at different frequencies and the composite signal, after captured and combined by the receiver, becomes distorted. Consequently, the dispersion in materials and hence different velocities should be taken into account in the design of a radar sensor, especially when the operating frequency range is large and the frequencies are high. Also, at frequencies above 1 GHz, water relaxation effect becomes dominant [1], making the dispersion more pronounced.

The attenuation of a signal traveling in a medium is determined by the medium's properties. Considering this and the fact that the properties of (simple) media are independent of the waveform type and sources, we can assume that the signal is a sinusoidal uniform plane wave to simplify the formulation without loss of generality. To further simply the analysis, we assume that there is no charge ($\rho = 0$) and current ($\vec{J} = 0$) in the medium, and the signal propagates into the medium along the direction z and represent the signal using only the electric and magnetic field components along the x and y directions, respectively. The (phasor) electric and magnetic fields can be determined from wave equations (2.5a, b) as

$$\vec{E} = \vec{a}_x E_o \mathrm{e}^{-\alpha z} \mathrm{e}^{-j\beta z} \tag{2.17}$$

and

$$\vec{H} = \vec{a}_y H_o \mathrm{e}^{-\alpha z} \mathrm{e}^{-j\beta z} = \vec{a}_y \frac{E_o}{|\eta|} \mathrm{e}^{-\alpha z} \mathrm{e}^{-j(\beta z - \phi_\eta)} \tag{2.18}$$

where E_o and H_o are the initial magnitudes of the electric and magnetic fields (at $z = 0$), respectively, with $\eta = |\eta| \angle \phi_\eta$ being the intrinsic impedance of the medium. The magnitudes of the electric and magnetic fields hence reduce exponentially according to e^{-az} as the signal propagates in the z direction.

Equations (2.17) and (2.18) show that the amplitudes of the electric and magnetic fields are attenuated by

$$A = \mathrm{e}^{-\alpha d} \tag{2.19}$$

or, in dB,

$$A_{dB} = -20\alpha d \log e = -8.686\alpha d \ (dB) \tag{2.20}$$

over a distance d.

The time-average power density ($\mathrm{W/m^2}$) of a signal is given as

$$\vec{S} = \frac{1}{2}\mathrm{Re}\left(\vec{E} \times \vec{H}^*\right) = \vec{a}_z \frac{E_o^2}{2|\eta|}\cos\left(\phi_\eta\right)\mathrm{e}^{-2\alpha z} \tag{2.21}$$

where (*) denotes the conjugate quantity. Equation (2.21) shows that power of a signal in a medium decreases exponentially at a rate of e^{-2az} as the signal travels in the direction of z. The total time-average power incident upon a surface S of a target can then be obtained as

$$\vec{P} = \frac{1}{2}\mathrm{Re}\int_S \left(\vec{E} \times \vec{H}^*\right) ds \tag{2.22}$$

Equations (2.21) and (2.22) allow the magnitude and direction of the real power of a signal travelling in any medium to be determined.

Some remarks need to be made here concerning the properties of materials involved in subsurface sensing and their general effects to the analysis and performance of a radar sensor. Subsurface media such as the asphalt and base materials of pavements are very complex. They are highly inhomogeneous, dispersive, and lossy. The inhomogeneity produces irregularities, which result in scattering and excessive high clutter noise, thus complicating the target detection. The dispersion distorts the return signal, and the high attenuation reduces and further distorts the return signal. As mentioned earlier, the properties of these materials are a function of frequency, space in the structures, material's constituents, and environmental factors. Furthermore, there is typically a close proximity between antennas and the ground encountered in the use of a radar sensor, and this complicates the radar signal's behavior. It is therefore very difficult, if not impossible, to analyze precisely the radar sensor's signal propagation in subsurface media existing in real testing environments. Consequently, this prevents an accurate calculation of the total loss of the signal as it propagates through the subsurface media and returns to the sensor.

2.3 Scattering of Signals Incident on Targets

Signals incident on a target are scattered and a portion of the scattered energy traveling toward the receiving antenna is captured by the antenna. In a given propagating medium, the scattering power is determined by the electromagnetic properties and physical structure of a target. As such, the power arriving to the receiving antenna can be estimated if the properties and structure of the target and propagating medium are known.

2.3.1 *Scattering of Signals on a Half-Space*

If a signal is incident on an interface between different media, part of its energy is reflected and part is transmitted through the interface. For simplicity, we assume the signal is a uniform plane wave. In the case of a flat surface, the reflection and transmission coefficients depend upon the polarization of the incident signal, the angles of incidence and transmission, and the intrinsic impedances of the media. There are two kinds of polarization: parallel and perpendicular polarizations. If the electric field is in the incident plane, the signal has a parallel polarization. Under the parallel polarization, and the magnetic field is perpendicular to the incident plane. Alternately, if the electric field is normal to the incident plane, the signal is in the perpendicular polarization and the magnetic field lies in the incident plane. Figure 2.1 illustrates the electric and magnetic fields for the parallel polarization.

2.3.1.1 Reflection at a Single Interface

Consider a single interface between two different media as shown in Fig. 2.1, by applying the boundary conditions for the tangential components of the electric and magnetic fields at the interface, we can derive the reflection (Γ_{par} and Γ_{per}) and transmission coefficients (T_{par} and T_{per}) for the parallel (par) and perpendicular (per) polarizations as

$$\Gamma_{par} = \frac{\eta_2 \cos\phi_t - \eta_1 \cos\phi_i}{\eta_2 \cos\phi_t + \eta_1 \cos\phi_i} \tag{2.23}$$

$$\Gamma_{per} = \frac{\eta_2 \cos\phi_i - \eta_1 \cos\phi_t}{\eta_2 \cos\phi_i + \eta_1 \cos\phi_t} \tag{2.24}$$

$$\mathrm{T}_{par} = \frac{2\eta_2 \cos\phi_i}{\eta_2 \cos\phi_t + \eta_1 \cos\phi_i} \tag{2.25}$$

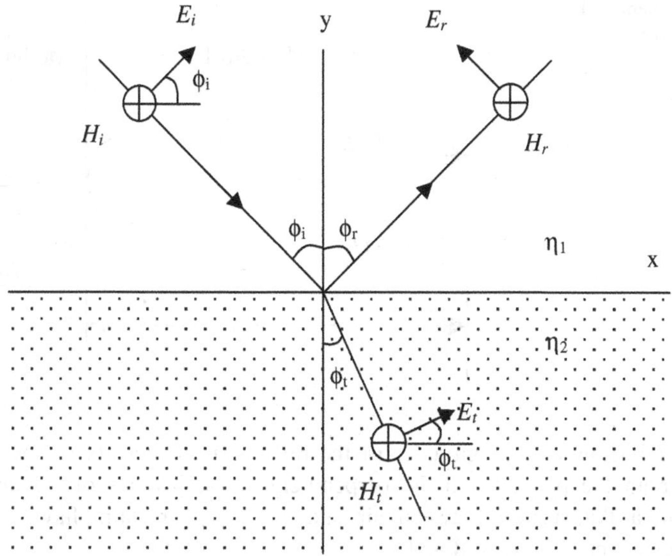

Fig. 2.1 Electric and magnetic fields of signals incident upon the interface between two different media for parallel polarization. The incident plane is *xy* plane

$$\mathrm{T}_{per} = \frac{2\eta_2 \cos \phi_i}{\eta_2 \cos \phi_i + \eta_1 \cos \phi_t} \tag{2.26}$$

respectively, where η_1 and η_2 are the intrinsic impedances of media 1 and 2, respectively, and ϕ_i and ϕ_t are the incident and transmitted angles, respectively. These angles are related by

$$\frac{\sin \phi_i}{\sin \phi_t} = \frac{\sqrt{\varepsilon_{r2}}}{\sqrt{\varepsilon_{r1}}} \tag{2.27}$$

where ε_{r1} and ε_{r2} are the relative dielectric constants of media 1 and 2, respectively. reflection and transmission coefficients are complex due to the fact that the intrinsic impedances are complex for (practical) lossy media.

Consider a parallel-polarized signal is incident on the interface through a lossy medium 1, as shown in Fig. 2.2, the time-average power density $S_r(R)$ reflected from the interface at the distance R from the interface can be found from (2.21) as

$$S_r(R) = \left|\Gamma_{par}\right|^2 \exp(-4\alpha_1 R) S_i(R) \tag{2.28}$$

where $S_i(R)$ is the average incident time-power density at R and α_1 is the attenuation constant of medium 1. The reflected power, which could be captured by the receiving antenna, is reduced due to the reflection at the interface and the attenuation in the propagating medium.

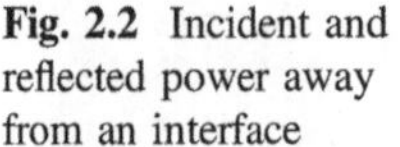

Fig. 2.2 Incident and reflected power away from an interface

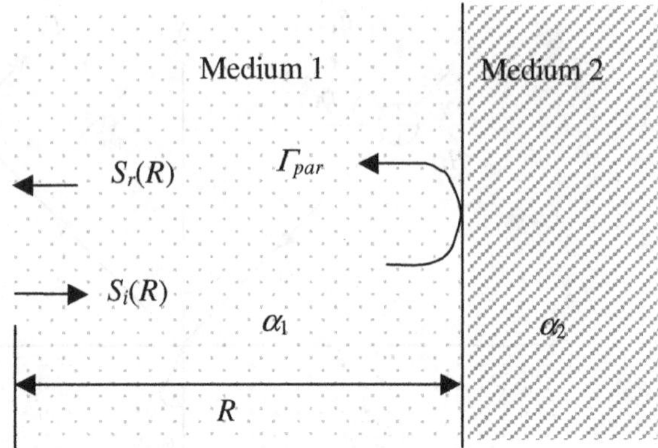

Figure 2.3 shows the magnitudes of the reflection coefficients versus incident angle for both the parallel and the perpendicular polarizations of a plane wave incident on a flat surface between air and a lossless medium having relative dielectric constant of 2, 4, 6, 8 and 10.

The phase of the reflected signal from an interface at a particular location, relative to that of the incident signal, is determined by the phase of the reflection coefficient, the velocity in the propagating medium, and the distance of the location from the interface.

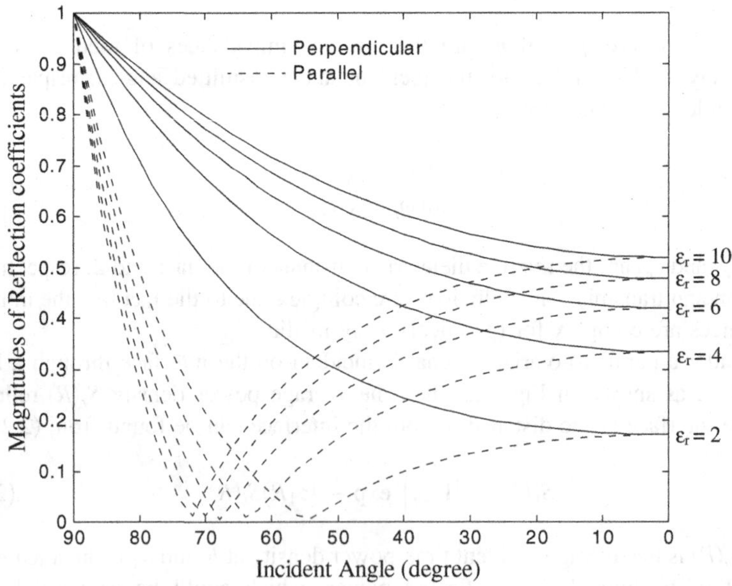

Fig. 2.3 Magnitudes of reflection coefficients of a plane wave incident on different dielectric materials from air

The reflection and transmission coefficients for normal incidence can be obtained from (2.23) to (2.26) by letting $\phi_i = 0$. For instance, the reflection coefficient for the parallel-polarization case is

$$\Gamma_{par} = \frac{\eta_2 - \eta_1}{\eta_2 + \eta_1} \tag{2.29}$$

For low loss and non-magnetic materials, such as those for pavements used in Table 2.1, the intrinsic impedance is almost real and hence can be approximated as $\eta = \sqrt{\mu_o/\varepsilon_o\varepsilon_r} = 377/\sqrt{\varepsilon_r}$ Ohms. Consequently, (2.29) can be rewritten as

$$\Gamma_{par} = \frac{\sqrt{\varepsilon_{r1}} - \sqrt{\varepsilon_{r2}}}{\sqrt{\varepsilon_{r1}} + \sqrt{\varepsilon_{r2}}} \tag{2.30}$$

where ε_{r1} and ε_{r2} are the real parts of the relative dielectric constants of media 1 and 2, respectively.

To verify possible use of (2.30) for practical pavement materials as shown in Table 2.1, the reflection coefficients at the interface between the asphaltand base layers are calculated using (2.29) and (2.30) as a function of the imaginary part of the relative permittivity of the base layer from 0.2 to 0.8 as shown in Table 2.1. As seen in Fig. 2.4, the reflection coefficient calculated from the approximate equation (2.30), assuming lossless materials, shows at most a 1 % difference from those determined from (2.29). Therefore, the assumption of lossless material is reasonable for calculating the reflection coefficients for the considered pavement materials. Similarly, the transmission coefficients can also be calculated for the considered pavement materials assuming lossless materials.

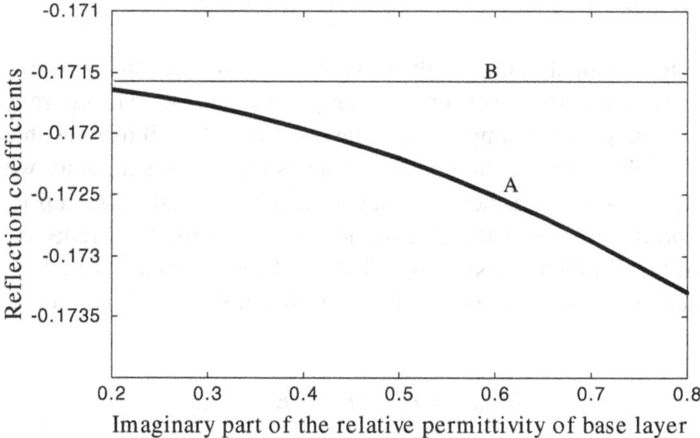

Fig. 2.4 Reflection coefficients A and B for normal incidence at the interface between asphalt and base calculated from (2.28) and (2.29), respectively

From (2.29) or (2.30), the phase of the reflection coefficient is either 0 or π radians. For instance, if the signal is incident from a material with lower dielectric constant to another with higher dielectric constant, the polarity of the reflected signal is opposite to that of the incident signal. This happens with most practical operations of subsurface-sensing radar sensors, such as those for used for assessment of pavements or detection of buried objects. On the contrary, when the incident signal propagates from a material with higher dielectric constant to one having lower dielectric constant, the reflected signal has the same polarity as the incident signal. The polarity of the reflected signal relative to that of the incident signal could be used to aid the analysis of a detected target. For instance, the identical polarity of the reflected signal as compared to the incident signal and the result can be used to detect an air void, which might indicate a defect in pavements, bridges, woods, walls, etc.

For highly lossy materials, where the lossless condition cannot be assumed, the resulting reflection coefficient is not close to real values. The complex reflection coefficient makes the phase of the reflected signal fall between 0 and 2π radians, depending on the losses of the materials. In addition, since the relative dielectric constant is a function of frequency, the phase of the reflection coefficient also changes with the frequency of the incident signal. Therefore, the relative dielectric constants of materials over the frequencies of interest are needed to determine the phases of the reflection coefficients.

2.3.1.2 Reflection and Transmission in Multi-layer Structures

The operations of radar sensors for sensing, particularly subsurface sensing such as pavement characterization, involve structures consisting of multiple layers. In order to analyze the signals propagating in involved media for extracting information of targets, the transmitted and reflected signals, and hence the transmission and reflection coefficients, at the interfaces between different layers need to be determined.

When a sign incident upon a multi-layer structure such as that shown in Fig. 2.5, reflection and transmission occur at multiple interfaces, causing reflected and transmitted signals propagating in these layers. Figure 2.5 illustrates these signals represented by their corresponding electric fields (E's). For simplicity without loss of generality, we only consider three layers and signal reflections up to the third interface. The magnitude of the electric field $E_{r,total}$ of the total reflected signal at the 1st interface, which consists of all the reflected signals, can be expressed approximately as the summation of all the electric fields of the individual reflected signals as

$$E_{r,total} = E_{r1} + E_{r2} + E_{r3} + E'_{r2} \tag{2.31}$$

where E_{r1} is the magnitude of the first reflected electric field at the first interface, E_{r2} is the magnitude of the transmitted electric field through the first interface from

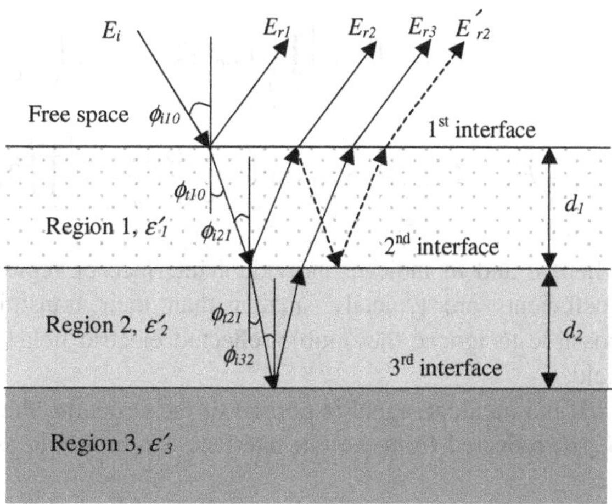

Fig. 2.5 Reflected and transmitted waves in a three-layer structure

the reflected signal at the second interface, E_{r3} is the magnitude of the transmitted electric field through the first interface from the reflected signal at the third interface, and E'_{r2} is the magnitude of the transmitted electric field through the first interface from another reflected signal at the second interface. E_{r1}, E_{r2} and E_{r3} are caused by a single reflection, corresponding to a single-reflected signal, and E'_{r2} is produced by double reflections, corresponding to a double-reflected signal, as can be seen in Fig. 2.5. These electric fields can be expressed as

$$E_{r1} = \Gamma_{10} E_i \tag{2.32}$$

$$E_{r2} = \mathrm{T}_{10}\Gamma_{21}\mathrm{T}_{01} E_i \exp\left(-\frac{2\alpha_1 d_1}{\cos\phi_{t10}}\right) E_i \tag{2.33}$$

$$E_{r3} = \mathrm{T}_{10}\mathrm{T}_{21}\Gamma_{32}\mathrm{T}_{12}\mathrm{T}_{01} \exp\left(-\frac{2\alpha_1 d_1}{\cos\phi_{t10}}\right) \exp\left(-\frac{2\alpha_2 d_2}{\cos\phi_{t21}}\right) E_i \tag{2.34}$$

$$E'_{r2} = \mathrm{T}_{10}\Gamma_{21}\Gamma_{01}\Gamma_{21}\mathrm{T}_{01} \exp\left(-\frac{\alpha_1 d_1}{\cos\phi_{t10}}\right) E_i \tag{2.35}$$

where Γ_{10} and T_{10} are the reflection and transmission coefficients of the signal incident from region 0 to region 1, respectively, ϕ_{t10} indicates the transmitted angle of the signal incident from region 0 to region 1, α_1 and α_2 are the attenuation constants of mediums 1 and 2, respectively, and d_1 and d_2 are the thicknesses of mediums 1 and 2, respectively. The single-reflected electric field magnitude E_{rn} at interface n and the double-reflected electric field magnitude in region n can be generalized, respectively, as

$$E_{rn} = \Gamma_{nn-1} \left[\prod_{m=1}^{n-1} \mathrm{T}_{mm-1} \mathrm{T}_{m-1m} \exp\left(-\frac{2\alpha_m d_m}{\cos \varphi_{tmm-1}} \right) \right] E_i \tag{2.36}$$

$$E'_{rn} = \Gamma^2_{nn-1} \Gamma_{n-2n-1} \exp\left(-\frac{2\alpha_n d_n}{\cos \varphi_{tnn-1}} \right) \left[\prod_{m=1}^{n-1} \mathrm{T}_{mm-1} \mathrm{T}_{m-1m} \right] E_i \tag{2.37}$$

where n and m indicate individual interface or region. In practice, the reflection coefficients are typically smaller than their transmission counterparts, making possible to ignore the double-reflected electric field in the total reflected electric field.

If the incident signal is normal to the structure, the time-average power density $S_{rn}(R)$ reflected from the nth interface can be found as

$$S_{rn}(R) = \Gamma^2_{nn-1} \left[\prod_{m=1}^{n-1} T^1_{mm-1} T^2_{m-1m} \right] \left[\prod_{k=1}^{n-1} \exp(-4\alpha_k d_k) \right] S_i(R) \tag{2.38}$$

where $S_i(R)$ is the time-average incident power density at R, α_k is the attenuation constant of the kth medium, and R is the distance from the interface. Equation (2.36) shows that the reflected power or returned power to the radar sensor will be significantly decreased if the transmission coefficients are small, as expected.

The phase of the transmitted signal propagating through an interface between two different materials, relative to that of the signal incident upon the interface, is determined by the phase of the transmission coefficient. As expressed in Eqs. (2.25) and (2.26), the magnitude of the transmission coefficient is real and positive, and the phase of the transmission coefficient can be between 0 and 2π radians. For lossless materials, especially, the transmission coefficient is a real value, which results in the transmitted signal having the same polarity as the incident signal. Note that the phase of the transmission coefficient depends upon the frequency of the incident signal as well as the losses of the materials.

2.3.2 *Radar Cross Section*

Radar Cross Section (RCS) of a target, which represents the back-scattering cross section of the target seen by a radar system (and hence is also named "backscatter cross section"), is an important parameter in the design and performance of radar sensors. The RCS mainly depends on the operating wavelength and angle from which the target is viewed by the system, and may be calculated or measured. The RCS is defined as the effective area of the target that intercepts the transmitted power and uniformly (or isotropically) radiates (scatters) all of the incident power in all directions [32]. It can be expressed as

Table 2.2 Radar cross sections of typical geometric shapes. λ is the wavelength

Geometric shapes	Dimension	RCS (σ)
Sphere	Radius r	πr^2
Flat plate	r × r	$4\pi r^2/\lambda^2$
Cylinder	H × radius r	$2\pi r H^2/\lambda$

$$\sigma = 4\pi \frac{\text{time-average scattered power at target toward receiver per unit solid angle}}{\text{time-average power density of incident wave at target}} \tag{2.39}$$

which is mathematically equivalent to

$$\sigma = \lim_{R\to\infty} 4\pi R^2 \frac{\left|\vec{E}_s\right|^2}{\left|\vec{E}_i\right|^2} = \lim_{R\to\infty} 4\pi R^2 \frac{P_s}{P_i} \tag{2.40}$$

where $\vec{E}_s$ and P_s are the scattered electric field and time-average power density magnitude at a distance R away from the target, respectively, $\vec{E}_i$ and P_i are the incident electric field and time-average incident power density magnitude at the target, respectively, and R is the distance between the target and receiving antenna (or the range). It can be inferred from (2.40) that the RCS is independent with R due to the fact that the power density is inversely proportional to R^2; this is expected since the RCS is a property of the target itself. The RCS provides system designers some crucial characteristics of targets observed by radar systems. When the range R is large with respect to wavelength, the incident signal is considered as a uniform plane wave.

Table 2.2 shows the theoretical RCS values of typical geometric shapes in optical regions (i.e., $2\pi r/\lambda > 10$) [23], where the ratio of the calculated RCS to the real cross sectional area of a sphere is 1. These values are very accurate as the RCS of a sphere is independent of the frequency in the optical region. The most typical geometry is a half-space for radar sensors used for surface or subsurface sensing involving structures of multiple media such as pavements consisting of asphalt, base and various subgrade layers, or walls in buildings. The half-space is considered to be an infinite plate that can be either smooth or rough according to the roughness of that plate [31].

2.4 System Equations

2.4.1 Friis Transmission Equation

Friis transmission equation, providing a very simple estimate of the received power with respect to the transmitted power for a general RF system, is a basic equation

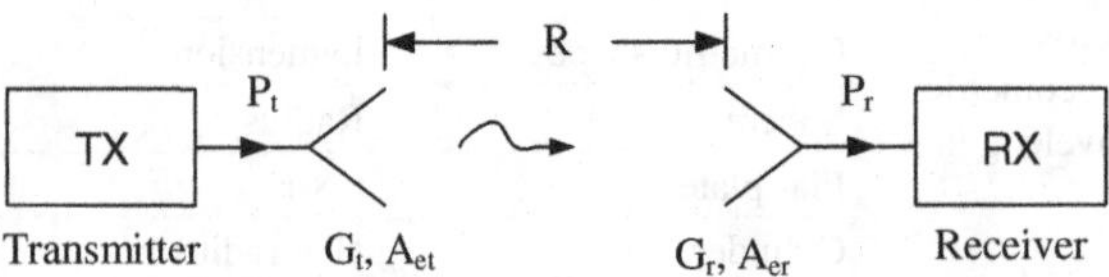

Fig. 2.6 Simple RF system's block diagram. P_t is the output power of the transmitter (TX), which is assumed to be equal to the power transmitted by the transmit antenna. P_r is the power arriving at receiver (RX), which is assumed to be equal to the power received by the receive antenna. G_t, A_{et} and G_r, A_{er} are the gain and effective antenna aperture of the transmit and receive antennas, respectively. R is the distance between the transmit and receive antennas

for communications and sensing. Figure 2.6 shows a simple (bi-static) RF system consisting of transmitter, receiver and antennas. For simplified illustration without loss of generality, we assume that the system and the transmission media are ideal—in which, the system has matched polarization; the antennas, transmitter and receiver are perfectly matched; the antennas are lossless; there is no scattering in signal's transmission and reception; and the antennas and transmission media are lossless.

The effective antenna area or aperture, which specifies the area of antenna that captures incoming energy, is defined as the ratio between the power received by the antenna and its power density. Mathematically, it can be derived as

$$A_e(\theta,\phi) = \frac{\lambda^2}{4\pi} G(\theta,\phi) \tag{2.41}$$

where θ and ϕ are the (angle) coordinates in a spherical coordinate system, λ is the operating wavelength, and $G(\theta,\phi)$ is the gain of the antennaAssume the transmit antenna is isotropic, the power density at the receive antenna, produced by the power illuminating from the transmit antenna, is given as

$$S_r = \frac{P_t}{4\pi R^2} \tag{2.42}$$

The receiving power density corresponding to a transmit antenna having gain G_t, is

$$S_r = \frac{P_t G_t}{4\pi R^2} \tag{2.43}$$

The power received by the receiver is given as

$$P_r = S_r A_{er} = \frac{P_t G_t}{4\pi R^2} A_{er} \tag{2.44}$$

We can derive, upon using (2.41) and (2.44),

$$\frac{P_r}{P_t} = G_t G_r \left(\frac{\lambda}{4\pi d}\right)^2 \tag{2.45}$$

which is known as the *Friis* transmission equation, which gives the optimum received power from a given transmitted power.

In practice, losses occur in system operations due to various reasons such as polarization mismatch between the transmit and receive antennas, scattering, mismatch loss at the transmitter and receiver, and loss in antennas and in the transmission medium. Taking these losses into account yields

$$\frac{P_r}{P_t} = G_t G_r L \left(\frac{\lambda}{4\pi R}\right)^2 \tag{2.46}$$

where $L < 1$ represents the total loss encountered by the system during operation including the system loss itself and loss of the propagating medium. The medium loss is accounted for by the loss factor $e^{-2\alpha R}$ for power, where α represents the attenuation constant of the medium.

The maximum range of detection or communication corresponds to the received power equal to the minimum power $P_{r,min}$, that can detected by the receiver, and can be determined from (2.46) as

$$R_{\max} = \frac{\lambda}{4\pi}\sqrt{G_t G_r L \frac{P_t}{P_{r,\min}}} \tag{2.47}$$

As can be seen, in order to double the range, the transmitting power must be increased four times, which is substantial.

2.4.2 Radar Equation

Radar equation governs the relationship between the transmitted and received power in radar systems, taking into account the systems' antenna gains, losses, operating frequencies, and ranges and radar cross sections of targets. It is an important equation used in the design and analysis of radar systems.

We consider a monostatic system using two separate antennas co-located or same antenna for transmitting and receiving, as shown in Fig. 2.7. The power density at the target is the same as that given in (2.43) with R denoting the distance from the antennas to the target as seen in Fig. 2.7. The power transmitted by the transmit antenna is intercepted and reradiated (scattered and reflected) by the target in different directions depending on the target's scattering characteristics. In the direction of the receive antenna, considering Fig. 2.7a, the reradiated power is derived as

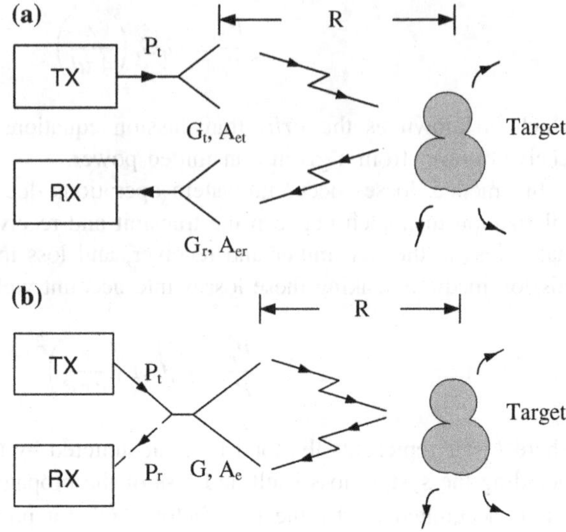

Fig. 2.7 A simple monostatic system with two antennas (**a**) and single antenna (**b**)

$$P_\sigma = \frac{P_t G_t}{4\pi R^2}\sigma \tag{2.48}$$

The power density at the receive antenna due to the power return from the target can be derived, under ideal conditions without any loss, as

$$S_r = \frac{P_t G_t \sigma}{(4\pi R)^2} \tag{2.49}$$

Using the antenna effective area as given in (2.41), the ratio between the power of the target's reflected signal received at the receiver and the power transmitted by the transmitter can be derived as

$$\frac{P_r}{P_t} = \sigma \frac{G_t G_r \lambda^2}{(4\pi)^3 R^4} \tag{2.50}$$

which is commonly known as the radar equation.

For practical systems operating under real conditions, the radar equation becomes

$$\frac{P_r}{P_t} = \sigma \frac{G_t G_r \lambda^2 L}{(4\pi)^3 R^4} \tag{2.51}$$

where $L < 1$ is again the total loss of the system under operation including the system loss and medium loss. The medium loss due to the propagating medium is

$e^{-4\alpha R}$ accounting for the propagation over twice of the target's range. The radar equation in (2.51) can be rewritten imposing this loss into it as

$$\frac{P_r}{P_t} = \sigma \frac{G_t G_r \lambda^2 L' e^{-4\alpha R}}{(4\pi)^3 R^4} \tag{2.52}$$

where L' represents the loss of the system itself. The radar equation involves the transmitted power, antenna gains, system loss and operating frequency, which are controlled and set by the system designer, and the target's RCS maximum range or penetration depth, and attenuation constant of the propagating medium, which, however, are not controlled by the system designer.

The maximum range can be determined from (2.51) and (2.52) as

$$R_{\max} = \left[\frac{P_t \sigma G_t G_r L \lambda^2}{(4\pi)^3 P_{r,\min}}\right]^{1/4} = e^{-\alpha R_{\max}} \left[\frac{P_t \sigma G_t G_r L' \lambda^2}{(4\pi)^3 P_{r,\min}}\right]^{1/4} \tag{2.53}$$

which is proportional to $P_t^{1/4}$. Note that the second expression for R_{max} in (2.53) represents a transcendental equation, which could be solved using a numerical method such as the Newton-Raphson method.

We can now see that, in order to double the maximum range, the transmit power needs to be increased by 16 times, which is very substantial and may not be achievable at RF frequencies for high-power and long-range applications using certain device technologies, particularly in the millimeter-wave regime. Equations (2.51) and (2.52) suggest that larger target's RCS leads to easier detection. As can be recognized by now, the RCS of objects, such as air planes or buried pipes, is a very important parameter to be considered in the design of objects and systems used to detect these objects.

When the transmit and receive antennas in Fig. 2.7a are the same, or considering a single antenna as shown in Fig. 2.7b, the radar equation becomes

$$\frac{P_r}{P_t} = \sigma \frac{G^2 \lambda^2 L}{(4\pi)^3 R^4} = \sigma \frac{G^2 \lambda^2 L' e^{-4\alpha R}}{(4\pi)^3 R^4} \tag{2.54}$$

where $G = G_t = G_r$ is the antenna gain, and the corresponding maximum range is

$$R_{\max} = \left[\frac{P_t \sigma L G^2 \lambda^2}{(4\pi)^3 P_{r,\min}}\right]^{1/4} = e^{-\alpha R_{\max}} \left[\frac{P_t \sigma L' G^2 \lambda^2}{(4\pi)^3 P_{r,\min}}\right]^{1/4} \tag{2.55}$$

Making use of (2.21), the squared magnitude of the incident field E_i, where $|E_i| = |E_0| e^{-\alpha R}$ with E_o being the initial amplitude of the transmitted electric field at the antenna, at the target can be obtained for a single-antenna monostatic system, as shown in Fig. 2.7b, as

$$|E_i|^2 = \frac{|2\eta|}{\cos\phi_\eta} \frac{P_t G e^{-2\alpha R}}{4\pi R^2} \tag{2.56}$$

The signal scattered from the target is attenuated as it propagates toward the antenna, and the squared magnitude of the scattered field E_s at the antenna can be derived as

$$|E_s|^2 = \frac{|2\eta|}{\cos\phi_\eta} \frac{P'_r}{A_e} \tag{2.57}$$

where $P'_r = P_r/G$ is the power captured by the antenna.

2.5 Signal-to-Noise Ratio of Systems

In practical operations, system's performance is affected by noise. Noise affecting a system operation can be classified into two kinds: external noise and internal noise. External noise represents noise caused by the environment surrounding the system, including noise injected from nearby stationary and moving objects. This noise is typically large at low frequencies but small in the RF range and is, in general, negligible as compared to the internal noise generated by the RF receiver itself. Receiver noise is the dominant noise in a system and is inherent in the receiver. In operation, the output signal of the receiver includes signals produced by desired targets as well as those from clutters, external noise and interference. Figure 2.8 shows a sketch of output voltage of a receiver. If the noise level contributed by the receiver itself is high or the received signal is weak, the system cannot perform accurately its intended function such as detecting a target. Typically, a threshold

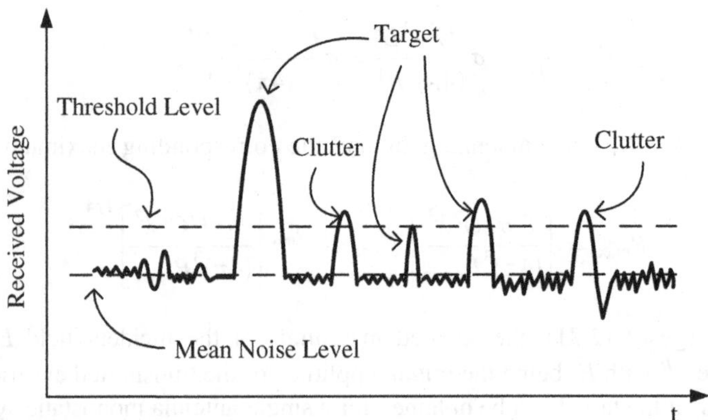

Fig. 2.8 Output voltage of a receiver

level is used to reduce the noise and clutter effects. However, if the threshold level is set to be sufficiently high, it will reduce the sensing capability of the system. For low threshold levels, on the other hand, inaccurate detection may result. Clutter effects can be reduced and identified by signal processing techniques. To increase the sensing capability or to enhance the communication performance of a system, the receiver's noise needs to be reduced or, equivalently, the receiver's signal-to-noise ratio (S/N) needs to be increased. This noise, although unavoidable, can be controlled to some extent by RF designers.

Receiver noise is, in general, contributed by three different noises. One is conversion noise generated during certain receiver operation—for example, FM-AM conversion noise. The other noise is low-frequency noise generated in the mixing process. The conversion and low-frequency noises depend on the receiver type—for instance, homodyne or FMCW receiver. The third noise contribution is thermal or Johnson noise generated by thermal motion of electrons in receiver's components. This noise always exists in receivers. We consider only thermal noise here.

The maximum thermal noise power available at the receiver's input is given as

$$P_{Ni} = kTB \tag{2.58}$$

where $k = 1.374 \times 10^{-23}$ J/K is the Boltzmann's constant, T is the temperature in Kelvin degree (K) at the receiver's input, and B is the noise bandwidth in Hertz (Hz), which is the absolute RF bandwidth over which the receiver operates. The available noise power P_{Ni} is independent of the receiver's operating frequency. In typically operating room temperature (62 °F), the thermal noise (kT) is about −174 dBm/Hz. As can be seen, this noise can be sufficiently large over a large bandwidth that degrades the noise performance of receiver and hence system substantially.

We define an ideal or noiseless receiver as a receiver that adds no additional noise as the input thermal noise P_{Ni} passes through it, except increasing the thermal noise level by the gain of the receiver. We now consider an actual (non-ideal) receiver that adds extra noise to that produced by an ideal receiver and define the noise figure of such receiver as

$$F = \frac{\textit{output noise power of actual receiver}\ P_{No}}{\textit{output noise power of ideal receiver}\ GP_{Ni}} = \frac{P_{No}}{kTBG} \tag{2.59}$$

where G is the available power gain of the receiver defined as

$$G = \frac{P_{So}}{P_{Si}} \tag{2.60}$$

with P_{Si} and P_{So} being the (real) signal's available power at the input and output of the receiver, respectively. The total noise power at the output of the receiver can then be obtained as

$$P_{No} = FkTBG = kT_nBG \tag{2.61}$$

where $T_n \equiv FT$, the thermal noise power per Hz, is defined as the (equivalent) "noise temperature" of the receiver. The noise figure of receivers can be obtained from (2.59) and (2.61) as the ratio between the input *S/N* and output *S/N* of receivers:

$$F = \frac{P_{Si}/P_{Ni}}{P_{So}/P_{No}} = \frac{\textit{signal-to-noise ratio at input}}{\textit{signal-to-noise ratio at output}} \tag{2.62}$$

which is more commonly known to RF engineers than (2.59). As can be seen, the noise figure indeed reduces the output S/N level of the receiver used in subsequent processing for sensing or communication purposes. This receiver's figure of merit contributes to the overall receiver's noise and is considered as one of the most important parameters of the receiver. In the design of receivers, it is important to minimize the noise figures of individual components, particularly those close to front of the receivers. A typical receiver consists of cascade of components—for example, a super-heterodyne receiver front-end primarily comprising of band-pass filter, low-noise amplifier, mixer and IF amplifier—and its noise figure depends on the receiver's individual components. It is noted that the noise figure of a passive component such as band-pass filter is equal to the reciprocal of the insertion loss of that component. For instance, a band-pass filter with a −3-dB insertion loss would have a noise figure equal to 2 or, in term of decibel, 3 dB. The (output) *S/N* of the receiver, making use of the radar equation (2.52), can be derived as

$$\frac{S}{N} = \sigma \frac{P_t G_t G_r \lambda^2 L' e^{-4\alpha R}}{(4\pi)^3 R^4 FkTB} \tag{2.63}$$

It is particularly noted that in actual system operations, the *S/N* value produced by receivers is much more important than the absolute powers of real and noise signals received by receivers.

2.6 Receiver Sensitivity

Receiver sensitivity indicates the minimum detectable input signal level for a receiver and hence measures the ability of a receiver to detect a signal. A system can detect a signal returned from a target or sent by another system if the received power is higher than the receiver sensitivity. The receiver sensitivity (*SR*) is determined by the noise temperature (*Tn*), bandwidth (*B*), noise figure (*F*), and signal-to-noise ratio (*S/N*) of the receiver.

Figure 2.9 depicts the sensitivity required for a receiver to detect a returned signal. The noise temperature (*kT*) entering the receiver is increased over the

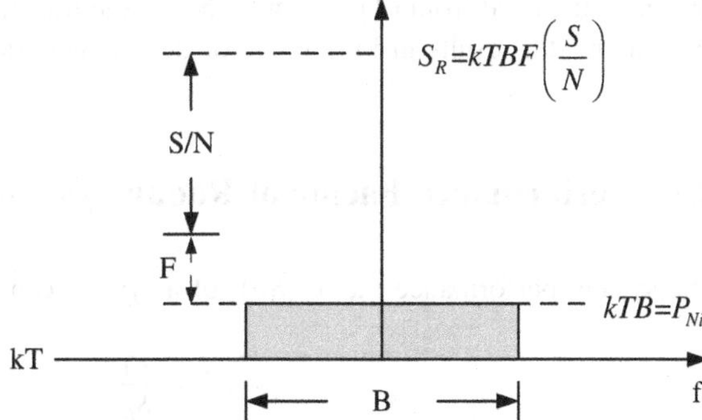

Fig. 2.9 Sensitivity of a receiver. $P_{Ni} = kTB$ is the input noise power

receiver's bandwidth to *kTB*, which is then further expanded through the noise figure and *S/N* of the receiver to reach a value of $kTBF(S/N)$. This final noise level is defined as the sensitivity of the receiver or the minimum input signal power that can be detected by the receiver:

$$S_R = kTBF\left(\frac{S}{N}\right) \tag{2.64}$$

Note that S_R is not the power density. The maximum range of a systems, achieved when the received power is equal to the receiver sensitivity, can be rewritten using (2.51) or (2.52) and (2.63) as

$$R_{\max} = \left[\frac{P_t \sigma G_t G_r L \lambda^2}{(4\pi)^3 kTBF\left(\frac{S}{N}\right)}\right]^{1/4} = e^{-\alpha R_{\max}}\left[\frac{P_t \sigma G_t G_r L' \lambda^2}{(4\pi)^3 kTBF\left(\frac{S}{N}\right)}\right]^{1/4} \tag{2.65}$$

A remark needs to be mentioned here concerning the average and peak powers. It is useful to consider the average power such as the average transmitting power, which is given as

$$P_{t,avg} = \frac{P_t}{B} \tag{2.66}$$

where *B* is the absolute (RF) bandwidth of the transmitter and receiver, in evaluating the system's parameter such as the maximum range. The average transmitting power is one of the controllable factors extensively used in designing a system and relates to the type of the waveform used. Using the average transmitting power, we can rewrite (2.65) as

$$R_{\max} = \left[\frac{P_{t,avg} \sigma G_t G_r L \lambda^2}{(4\pi)^3 kTF\left(\frac{S}{N}\right)}\right]^{1/4} = e^{-\alpha R_{\max}}\left[\frac{P_{t,avg} \sigma G_t G_r L' \lambda^2}{(4\pi)^3 kTF\left(\frac{S}{N}\right)}\right]^{1/4} \tag{2.67}$$

It can be deduced from (2.67) that high average transmitting power combined with less bandwidth results in long range or deep penetration.

2.7 Performance Factor of Radar Systems

The system performance factor (SF) of a system can be defined as [33]

$$SF = \frac{P_t}{S_R} \tag{2.68}$$

This factor is the system's figure of merit used to measure the overall performance of a system and is one of the most important parameters in the system equation for estimating the system's range. As the minimum detectable signal corresponds to the maximum range of a system, we can derive the system performance factor, utilizing (2.51) or (2.52), (2.64) and (2.68), by letting the received power equal to the receiver sensitivity (S_R) as

$$SF = \frac{(4\pi)^3 R_{\max}^4}{G_t G_r L \sigma \lambda^2} = \frac{(4\pi)^3 R_{\max}^4}{G_t G_r L' \sigma \lambda^2 e^{-4\alpha R_{\max}}} \tag{2.69}$$

The performance factor given in (2.69) neglects the contribution of the receiver. In practical systems, however, the system performance factor is limited by the actual receiver dynamic range. Hence, it is necessary to incorporate a correction for the receiver dynamic range into the system performance factor as we will address in the analysis of SFCW radar sensor in Chap. 3.

2.8 Radar Equation and System Performance Factor for Targets Involving Half-Spaces

The radar equations described in Sect. 2.5 are general radar equations. To provide more details and insight for specific applications such as those involving sensing of multi-layer structures like pavements or buried objects, the radar equation needs to be modified taking into account the specific conditions encountered in such applications to enable more accurate characterization such as estimation of maximum range or penetration depth.

We first consider a single half-space target spaced at a distance R from an antenna as shown in Fig. 2.10a and assume a uniform plane wave incident normal to the interface. Following the derivation of (2.44), we can derive the power received by the antenna, considering the reflection at the interface, the roundtrip travel of $2R$, and the loss of the propagating medium, as

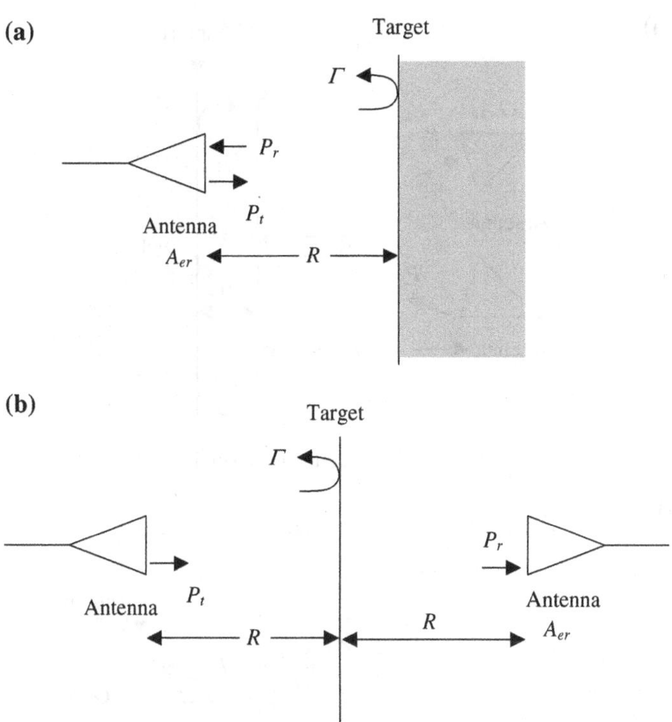

Fig. 2.10 Single half-space target illuminated with a uniform plane wave from an antenna (**a**) and its equivalent using the image technique (**b**)

$$P_r = \frac{P_t G_t A_{er} L}{4\pi (2R)^2} \Gamma^2 \tag{2.70}$$

where Γ is the magnitude of the reflection coefficient and L (<1) is the loss of the medium. The *RCS* of a half-space is found from (2.51) utilizing (2.41) and (2.70) as

$$\sigma = \pi R^2 \Gamma^2 \tag{2.71}$$

Equation (2.39) can also be derived considering Fig. 2.10b obtained by applying the image technique presented in [31].

We now consider a target consisting of two layers with each layer assumed to be a half-space to exemplify the analysis of a more general multi-layer target. This simple structure simplifies the formulation and show signal interactions without loss of generality. Figure 2.11a shows a two-layer half-space target with a uniform plane wave traveling obliquely to the first interface and the antennas with the same gain are located next to each other. Extending the foregoing analysis for a single interface to two consecutive interfaces depicted in Fig. 2.11a or analyzing the

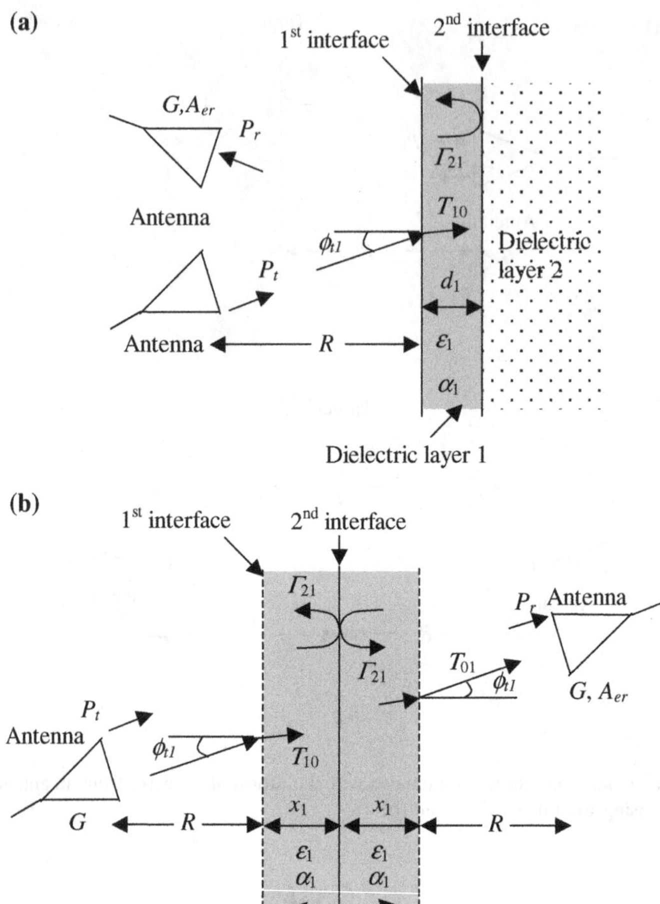

Fig. 2.11 Two-layer half-space target illuminated with a uniform plane wave (**a**) and its equivalent using the image technique (**b**)

equivalent structure shown in Fig. 2.11b obtained by applying the image principle, we can derive the power received by the antenna from (2.70). To that end, we replace the reflection coefficient Γ in (2.70) with the composite reflection coefficient upon reflections from the two interfaces from (2.33) as

$$\Gamma = \mathrm{T}_{10}\Gamma_{21}\mathrm{T}_{01}\exp\left(-\frac{2\alpha_1 d_1}{\cos\phi_{t10}}\right) \tag{2.72}$$

and the distance R in Eq. (2.70) with, considering the oblique incidence,

$$R = \frac{R}{\cos\phi_{i1}} + \frac{x_1}{\cos\phi_{t10}} \tag{2.73}$$

where ϕ_{i1} and ϕ_{t10} are the incident and transmitted angles at the first interface, respectively. Note that the thickness of layer 1, d_1, should be replaced with $x_1 = d_1\sqrt{\varepsilon_{r1}}$, where ε_{r1} is the relative dielectric constant of layer 1, as the signal's velocity is reduced in the layer. The radar equation, which determines the power received at the antenna upon reflection from the second interface, can now be expressed as

$$P_{r2} = \frac{P_t G^2 \lambda^2 L}{(4\pi)^2 \left(\frac{2R}{\cos\phi_{i1}} + \frac{2x_1}{\cos\phi_{t10}}\right)^2} \Gamma_{21}^2 \mathrm{T}_{10}^2 \mathrm{T}_{01}^2 \exp\left(-\frac{4\alpha_1 d_1}{\cos\phi_{t10}}\right) \tag{2.74}$$

The result in (2.74) can be generalized to obtain the radar equation for multi-layer targets with each layer assumed to be a half-space as

$$P_{rn} = \frac{P_t G^2 \lambda^2 L \Gamma_{nn-1}^2 \left[\prod_{m=1}^{n-1} \mathrm{T}_{mm-1}^2 \mathrm{T}_{m-1m}^2 \exp\left(\frac{-4\alpha_m d_m}{\cos\phi_{tmm-1}}\right)\right]}{(4\pi)^2 \left(\frac{2R}{\cos\phi_{i1}} + \sum_{l=1}^{n-1} \frac{2x_l}{\cos\phi_{tll-1}}\right)^2} \tag{2.75}$$

where P_{rn} is the power arriving at the receive antenna from the nth interface.

The nth interface is detectable if $P_{rn} \geq S_R$. Consequently, the radar's system performance factor SF is found using (2.68), (2.69) and (2.75) as

$$SF = \frac{64\pi^2 \left(\frac{R}{\cos\phi_{i1}} + \sum_{l=1}^{n-1} \frac{x_l}{\cos\phi_{tll-1}}\right)^2}{G^2 \lambda^2 L' \Gamma_{nn-1}^2 \left[\prod_{m=1}^{n-1} \mathrm{T}_{mm-1}^2 \mathrm{T}_{m-1m}^2 \exp\left(-\frac{4\alpha_m d_m}{\cos\phi_{tmm-1}}\right)\right]} \tag{2.76}$$

Equation (2.76) can be used to estimate the maximum range or penetration depth of radar sensors for sensing multi-layer half-space targets such as pavement layers.

2.9 Radar Equation and System Performance Factor for Buried Objects

Figure 2.12 shows an object buried in a medium underneath a surface, which is illuminated with a uniform plane wave in air (assumed to be lossless) from an antenna. The time-average power density S at the object can be derived from (2.43), taking into account the oblique incidence, transmission coefficient T_{10} at the surface and attenuation constant α_1 of the medium, as

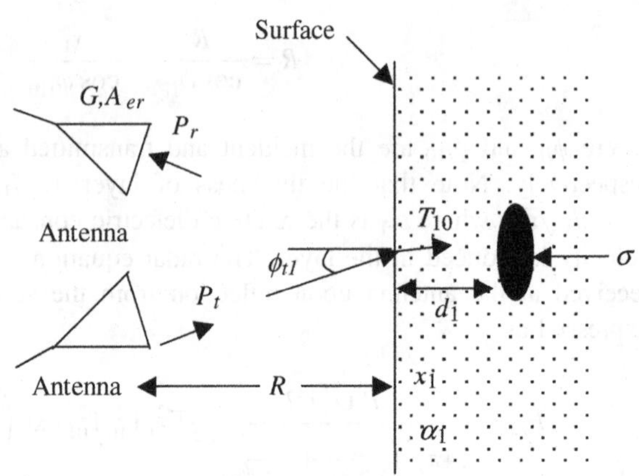

Fig. 2.12 Buried object under a surface

$$S = \frac{P_t G_t}{4\pi\left(\frac{R}{\cos\varphi_{i1}} + \frac{x_1}{\cos\varphi_{t10}}\right)^2} \mathrm{T}_{10}^2 \exp\left(-\frac{2\alpha_1 d_1}{\cos\phi_{t10}}\right) \tag{2.77}$$

The *RCS* of the object is assumed to be approximately equal to that of a half-space as

$$\sigma = \pi\left(\frac{R}{\cos\phi_{i1}} + \frac{x_{1\max}}{\cos\phi_{t10}}\right)^2 \Gamma^2 \tag{2.78}$$

where Γ is the reflection coefficient at the object's surface. The reflected power from the object can then be obtained from (2.74) making use of (2.41) and (2.71) as

$$P_r = \frac{P_t G A_{er} \sigma L'}{(4\pi)^2\left(\frac{R}{\cos\phi_{i1}} + \frac{x_1}{\cos\phi_{t10}}\right)^4} \mathrm{T}_{10}^2 \mathrm{T}_{01}^2 \exp\left(-\frac{4\alpha_1 d_1}{\cos\phi_{t10}}\right) \tag{2.79}$$

Consequently, the system performance factor defined in (2.68) can be derived as

$$SF = \frac{64\pi^3\left(\frac{R}{\cos\phi_{i1}} + \frac{x_{1\max}}{\cos\phi_{t10}}\right)^4}{G^2\lambda^2\sigma \mathrm{T}_{10}^2 \mathrm{T}_{01}^2 L' \exp\left(-\frac{4\alpha_1 d_{1\max}}{\cos\phi_{t10}}\right)} \tag{2.80}$$

where $d_{1\max} = x_{1\max}/\sqrt{\varepsilon_{r1}}$, from which the maximum detectable range, d_{1max}, under the surface can be determined.

2.10 Targets Consisting of Multiple Layers and Buried Objects

The analysis becomes more complex for targets involving objects buried within multi-layer structures such as those shown in Fig. 2.13. This target model illustrates some practical targets such as a pavement or wood composite consisting of multiple layers with defects such as voids. Figure 2.13 shows multiple transmissions and reflections, represented by *T*'s and *Γ*'s, respectively, resulted from a signal incident from air upon the surface. Conventional techniques of modeling wave propagation in a multi-layer structure that lead to the radar equations and system performance factors, as described in the foregoing Sects. 2.8 and 2.9, are based on the following assumptions: (1) only a propagating uniform plane wave exists in the entire structure, (2) the layers are homogeneous (e.g., no voids), and (3) reflection coefficient of the final wave reflected off of the surface is the sum of all the reflection coefficients of the individual waves reflected pass the first interface toward the left. These assumptions are not very accurate for structures having objects embedded in multiple layers such as that shown in Fig. 2.13. When a signal is incident this structure at high frequencies, the incident wave will excite an infinite number of different waves, including propagating and evanescent waves, at the first interface. These generated waves are reflected and transmitted into air and the first layer, respectively. Part of the transmitted waves will also produce other reflected and transmitted waves upon incident to the buried objects in the first layer. This process will then continue in subsequent layers. Although the evanescent waves die with distance away from the interfaces and buried objects, their effects would be significant near the interfaces and objects at these frequencies and must be considered for accurate determination of the final reflection coefficient and hence the radar equation and system performance factor. These effects can be considered in the propagation analysis using various full-wave EM techniques such as the mode-matching method [34].

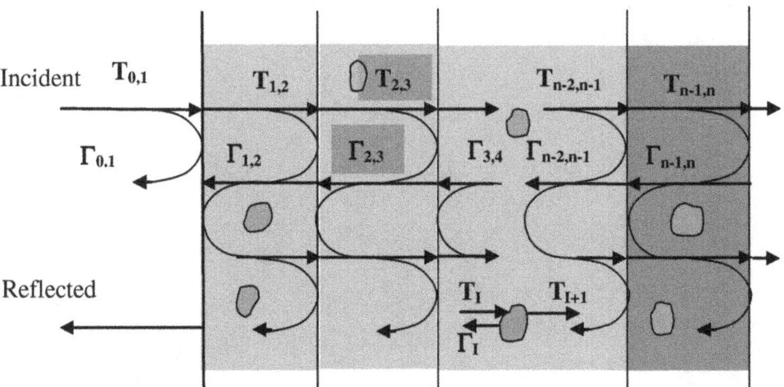

Fig. 2.13 Buried objects within multiple layers

2.11 Summary

This chapter addresses the general analysis of radar sensors. Specifically, it discusses signal propagation in media encountered in using radar sensors, signal scattering from objects, systems equations including Friis transmission equation and radar equations, signal-to-noise ratio, receiver sensitivity, maximum range or penetration depth, and system performance factor. Understanding these basic parameters provides general insight of a radar sensor's possible performance and operation and allows its general analysis to be conducted, which are essential toward the design of microwave systems for sensing applications.

Chapter 3
Stepped-Frequency Radar Sensor Analysis

3.1 Introduction

This chapter addresses various topics concerning the analysis of stepped-frequency continuous-wave (SFCW) radar sensors including their principles, which contains discussions of transmitted and received signals, down-converted digitized in-phase (I) and quadrature-phase (Q) signals and their complex I/Q vectors, and synthetic pulses characterizing targets; design parameters, comprising angle and range resolutions, frequency step, number of frequency steps, total bandwidth, range or penetration depth, range accuracy, range ambiguity, and pulse repetition interval; dynamic range and system performance factor; and estimations of the maximum ranges or penetration depths in sensing multi-layer and buried targets.

3.2 Principle of Stepped-Frequency Radar Sensors

SFCW radar sensors transmit consecutive trains of sinusoidal signals at different frequencies toward targets, receive the reflected signals from the targets, and process them for the targets' information. In this process, the received signals at each stepped frequency are down-converted to an intermediate-frequency (IF) signal, which is then demodulated into base-band in-phase (I) and quadrature (Q) signals. The I and Q signals contain both the amplitude and phase information of the targets. These signals are converted into digital signals with the aid of an analog-to-digital converter (ADC) and then transformed into a "synthetic pulse" in time domain using Inverse Discrete Fourier Transform (IDFT). The synthetic pulse represents the target electrically and is processed to reveal the target's characteristics. It is noted that the amplitudes and phases of the transmitted sinusoidal signals are typically not weighted to achieve particular amplitudes and phases. While properly weighting the amplitudes and phases of the transmitted signals would improve the

C. Nguyen and J. Park, *Stepped-Frequency Radar Sensors*,
SpringerBriefs in Electrical and Computer Engineering,
DOI 10.1007/978-3-319-12271-7_3

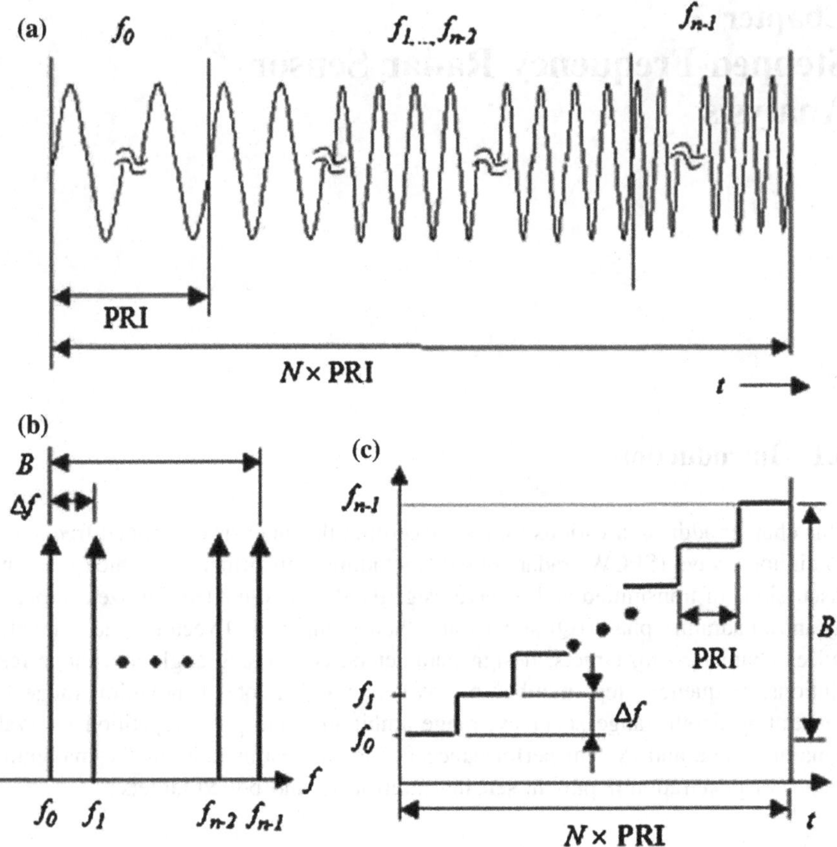

Fig. 3.1 Transmitted signals and frequencies of SFCW radar sensors: **a** waveforms, **b** frequencies, and **c** frequencies versus time

performance of SFCW radar sensors, it is very difficult in practice to design a microwave transmitter that can transmit signals at different frequencies with specified amplitudes and phases, especially in the millimeter-wave region. Detailed analysis of this procedure of SFCW radar sensors is given as follows.

The transmitted frequencies (f_0, f_1, …, f_{N-1}) are separated with a uniform frequency step Δf, as shown in Fig. 3.1, that depicts the transmitted signals and their frequencies. SFCW radar sensors are essentially operated as frequency-modulation systems with their frequencies modulated as shown in Fig. 3.1c. As can be inferred from Fig. 3.1, the total bandwidth B is $N\Delta f$, where N is the number of the frequency steps. *PRI* is the pulse repetition interval of the synthesizer of the transmitter, defined as the time required for transmitting each constituent signal of a single frequency, which is an important parameter needed to be considered in SFCW radar sensors.

In operation, a SFCW radar sensor illuminates each individual target with all signals across the entire operating frequency range; hence the speed of the transmitter's synthesizer in transmitting the entire spectrum needs to be sufficiently fast and the synthesizer's settling time needs to be as short as possible. For stationary sensing, where a SFCW radar sensor is fixed in position, the synthesizer's speed does not really affect the operation of the sensor provided that the time that the system takes to display the results is acceptable. However, in non-stationary sensing applications such as continuous characterization of actual roads or sensing across large underground areas, where the sensor is placed on a moving platform like vehicle or aircraft, the synthesizer's speed needs to be sufficiently fast to accommodate the speed of the platform. The faster the speed of the platform, the faster the synthesizer's speed is required. For instance, the synthesizer's speed needs to be less than 25 ms if the platform moves at 20 miles per hour.

Moreover, as can be inferred from the analyses in this section and subsequent sections, complete transmission of the signals at all frequencies is important for SFCW radar sensors to function properly such as producing an accurate construction of the synthetic pulse from a target for extracting the target's information. Missing the transmission of a signal at any frequency would degrade the sensor's performance such as causing errors in the generation of targets' synthetic pulses, which in turn result in undesired effects such as increase of self-clutter of the reconstruction leading to difficulty in extracting and less accurate sensing results. To that end, the synthesizer needs to function properly at all time and the antenna needs to be designed to cover all the desired frequencies.

It should also be mentioned that the frequency instability or phase noise of the synthesizer contributes to the errors that affect measurement accuracy. The phase noise of the synthesizer's signal is down-converted and appears in the IF signal. The stability requirement of the synthesizer can be predicted corresponding to the time delay between the transmit and receive signals [35].

An important remark for SFCW radar sensors, which is also significant for other radar sensors, is the isolation between the transmitter and receiver, especially when a single antenna is used for both transmission and reception along with a circulator or transmit/receive (T/R) switch. This isolation is especially crucial in order to reduce or prevent the RF leakage from the transmitter to receiver under the reception mode. Undesired RF leakage from the transmitter adds to the received signal, causing detrimental effects such as saturating the receiver, distorting the received signal, and reduced the system's dynamic range and linearity. Moreover, the received signal from the antenna, after being amplified by the low-noise amplifier (LNA) in the receiver, can leak to the transmitting path, amplified by the power amplifier (PA) of the transmitter, and then goes back to the antenna. This leaking signal may be larger than the received signal, hence distorting the received signal at the antenna. To resolve these problems, the PA (and some other components) in the transmitting path are typically turned off during the reception mode. This results in some operation and performance issues. For instance, as the PA and other components are switched on and off frequently, they require some time in each state to completely settle before the transmission or reception process starts, inadvertently slowing down the system's operating speed and reducing the component lives. A better solution to overcome the RF leakage problem is to utilize an ultra-high isolation T/R switch, leading to overall improvement in noise figure, dynamic range and linearity while enabling the system to continuously operate with all the transmitter components being on at all time. Such an ultra-high isolation T/R switch can be designed by implementing the RF leakage cancellation technique described in [36].

Another important remark is, like for other radar sensors, the synthesizer and other non-linear components of the transmitters such as power amplifiers need to be designed to suppress the harmonics of the fundamental signals and intermodulation products as much as possible, and proper filtering needs to be incorporated into the system, whether through a band-pass filter following the antenna, the antenna itself, or other components after the antenna like a T/R switch. Achieving these suppressions would reduce unwanted RF radiation, overcoming potential problems of false targets and leading to less interference with other RF operations, RF environment pollution, and stealthy operations, etc.

The transmitted waveforms of SFCW radar sensors can be expressed as [23]

$$x_i(\omega_i, t) = A_i \cos[\omega_i t + \theta_i] \tag{3.1}$$

where $\omega_i = 2\pi(f_0 + i\Delta f)$, $i = 0, \ldots, N - 1$, is the ith angular frequency and A_i and θ_i are the amplitude and the relative phase of the ith transmitted signal, respectively. The signals returned to the receiver upon reflections from a stationary target can be represented as

$$r_i(\omega_i, t, \tau) = B_i \cos[\omega_i(t - \tau) + \theta_i] \tag{3.2}$$

where B_i is the amplitude of the ith returned signal and τ is the two-way travel time between the sensor and target, which is directly related to the range R of the target as

$$R = \frac{v\tau}{2} \tag{3.3}$$

where v is the speed of signals in the propagating medium—for air, it is $c = 3 \times 10^8$ m/s.

The normalized base-band I and Q signals (with unit amplitude), obtained by demodulating the returned signals from the target, can be written as

$$I_i(\omega_i, \tau) = \cos(-\omega_i \tau) = \cos\phi_i \tag{3.4a}$$

$$Q_i(\omega_i, \tau) = \sin(-\omega_i \tau) = -\sin\phi_i \tag{3.4b}$$

which are sampled into digital I and Q signals through an ADC. It is noted that the range information of the target can be found from the phase $\phi_i = \omega_i\tau$ of the I and Q signals. Upon digitizing these I and Q components, the digitized I and Q signals are combined into a complex vector as

$$C_i(\omega_i, \tau) = I_i(\omega_i, \tau) + jQ_i(\omega_i, \tau) = \exp(-j\omega_i\tau) = \exp(-j\phi_i) \tag{3.5}$$

from which a complex vector array V corresponding to a sequence of N stepped-frequencies can be formed as

$$V = [C_0, C_1, \ldots, C_{N-1}] \tag{3.6}$$

Applying the IDFT on the complex vector C_i transforms this vector in the frequency domain into a range profile of the target in the time domain as [23]

$$y_n = \frac{1}{M}\sum_{i=0}^{M-1} C_i \exp\left[\frac{j2\pi ni}{M}\right] \tag{3.7}$$

where $0 \leq n \leq M - 1$. Adding $(M - N)$ zeros into the array V to make its size into the power of two increases the speed of the IDFT and the target's range accuracy. This results in a new sequence of the array V_k consisting of M vectors, where $k = 1, 2, \ldots, M - 1$. Applying the IDFT to the array V_k then gives

$$y_n = \frac{1}{M}\sum_{k=0}^{M-1} V_k \exp\left[\frac{j2\pi nk}{M}\right] \tag{3.8}$$

where $V_k = \exp(-j\phi_k) = \exp(-j\omega_k\tau)$ for $1 \leq k \leq N - 1$ and zero otherwise. Substituting V_k into (3.8) gives

$$y_n = \frac{1}{M}\sum_{k=0}^{M-1} \exp\left[j\left(\frac{2\pi nk}{M} - \phi_k\right)\right] \tag{3.9}$$

where

$$\phi_k = \phi_{\left(k - \frac{M-N}{2}\right)} \tag{3.10}$$

Equation (3.10) states that ϕ_k is valid only if $(M - N)/2 \leq k \leq (M + N)/2 - 1$, otherwise $\phi_k = 0$. Equation (3.9) becomes, letting $k = m + (M - N)/2$:

$$y_n = \frac{1}{M}\sum_{m=0}^{N-1} \exp\left[j\left(\frac{2\pi n}{M}\left(m + \frac{M-N}{2}\right) - \phi_m\right)\right] \tag{3.11}$$

Equation (3.11) can be rewritten in terms of the target's range R, after normalizing, as

$$y_n = \sum_{m=0}^{N-1} \exp\left[j\left(\frac{2\pi n}{M}\left(m + \frac{M-N}{2}\right) - \frac{2\pi f_m 2R}{c}\right)\right] \tag{3.12}$$

which, upon rearrangement, becomes

$$y_n = \exp\left(j\frac{4\pi f_0 R}{c}\right)\exp\left(j\frac{\pi n(M-N)}{M}\right)\sum_{m=0}^{N-1}\exp\left[j\left(\frac{2\pi n}{M} - \frac{2\pi\Delta f 2R}{c}\right)m\right] \tag{3.13}$$

where $f_m = f_0 + m\Delta f$ with f_0 being the start frequency. Solving (3.13) gives

$$y_n = \exp\left(j\frac{4\pi f_0 R}{c}\right)\exp\left[j\frac{\pi n(M-N)}{M}\right]\exp\left[j\frac{a(N-1)}{2}\right]\frac{\sin\left(\frac{aN}{2}\right)}{\sin\left(\frac{a}{2}\right)} \tag{3.14}$$

where $a = \left(n - \frac{2M\Delta fR}{c}\right)\frac{2\pi}{M}$. Taking the magnitude of (3.14) gives

$$|y_n| = \left|\frac{\sin\left(\frac{aN}{2}\right)}{\sin\left(\frac{a}{2}\right)}\right| \tag{3.15}$$

where N is again the number of frequency steps. Equation (3.15) represents essentially the target response of SFCW radar sensors (upon performing the IDFT), which contains the target's information sought by the radar sensors. This response has a shape of pulses and hence is referred to as the "synthetic pulse" of SFCW radar sensors.

Figure 3.2 shows a synthetic pulse of SFCW radar sensors where the pulse, consisting of N lobes, is repeated every M cells due to M points of IDFT [33], where M is the number of IDFT. The peaks of the main lobes of the synthetic pulse obtained from (3.15) occur when $n = n_p + lM$ corresponding to $a = \pm 2l\pi$, $l = 0, 1, 2, \ldots$, where n_p is the cell number corresponding to the peak of the synthetic pulse's main lobe at $a = 0$. Hence, the range of the target in terms of n_p becomes

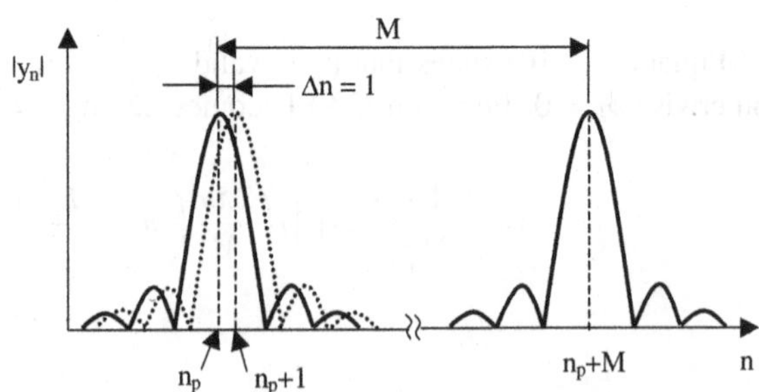

Fig. 3.2 Synthetic pulse of SFCW radar sensors

$$R = \frac{n_p v}{2M\Delta f} \tag{3.16}$$

The minimum displacement of the peak of the main lobe range, as shown in Fig. 3.2, is the minimum displacement of range (or range deviation) corresponding to one cell, $\Delta n = 1$, and represents the range accuracy. It can be obtained from (3.16) by letting $n_p = 1$ as

$$\delta R = \frac{v}{2M\Delta f} \tag{3.17}$$

It is important to note that the range given by (3.16) and the range accuracy from (3.17) are valid without error if the frequency source, which supplies the transmitted signals, is ideal and the frequency step-size Δf is uniform. In practice, however, sources are not ideal and hence their signals are contaminated with noises, and the frequency step is non-uniform. Consequently, these cause effects on the range measured and limit the range accuracy, beyond which it cannot be improved. Specifically, the measured results would be slightly different from the values obtained by (3.16) and (3.17).

The (un-normalized) complex vector of the *I*/*Q* components is obtained from (3.5) as

$$C_i(\omega_i, \tau) = I_i(\omega_i, \tau) + jQ_i(\omega_i, \tau) = A_i \exp(-j\phi_i) \tag{3.18}$$

where A_i the amplitude of the base-band *I* and *Q* signals. When the SFCW radar sensors receive a train of stepped frequencies from a stationary point target at range *R*, the phase of the complex *I*/Q vectors is

$$\phi_i = -2\pi f_i \tau = -\frac{4\pi f_i R}{v} \tag{3.19}$$

where f_i is the *i*th frequency and τ is again the two-way travel time. The change of the phase ϕ_i of the complex *I*/Q vectors, with respect to time, produces a constant radian frequency of

$$\omega = -\frac{\partial \phi_i}{\partial t} = -\frac{4\pi R}{v}\frac{df_i}{dt} = -\frac{4\pi R \Delta f}{v(PRI)} \tag{3.20}$$

If the target range *R* is fixed, the complex *I*/Q vectors with a magnitude of A_i (assumed to be constant) rotate at a constant rate along the locus as shown in Fig. 3.3, where the phase ϕ_i is a function of the stepped-frequency f_i as seen in (3.20).

The above analysis can be generalized for multiple targets. For simple illustration, we consider two targets at ranges R_1 and R_2. The magnitudes of the complex vectors for the targets at R_1 and R_2 are A_i and B_i, respectively. For simplicity, we

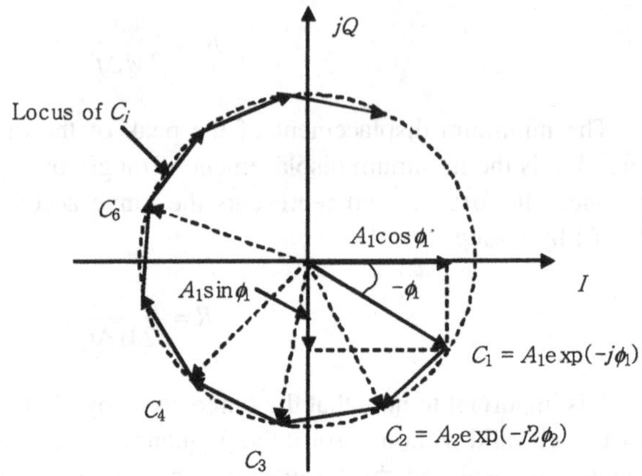

Fig. 3.3 Complex I/Q vectors for a fixed point target rotating at a constant rate along the locus with the amplitudes A_i of the returned signals assumed to be constant

assume the phases for these vectors are identical as ϕ_i. We also assume that the radian frequency ω of the composite complex I/Q vectors is the sum of the individual radian frequencies (ω_{R1} and ω_{R2}) of the complex I/Q vectors produced by the two targets individually. This frequency is expressed as

$$\omega = \omega_{R1} + \omega_{R2} = -K(R_1 + R_2) \tag{3.21}$$

where $K = -4\pi\Delta f/[v(PRI)]$ is constant if the targets are located in the same propagating medium. The vector diagram resulting from the two targets is depicted in Fig. 3.4.

3.3 Design Parameters of Stepped-Frequency Radar Sensors

The design of SFCW radar sensors involves various design parameters including resolution, frequency step, number of frequency steps, total bandwidth, range or penetration depth, range accuracy, range ambiguity, and pulse repetition interval.

The resolution of a radar sensor can be either range or angle resolution, depending on the direction of observation. The range resolution depends upon the total absolute bandwidth of the transmitted signals and their velocity. On the other hand, the angle resolution is directly proportional to the 3-dB beam-width of the antenna and the distance between the antenna and the target. Therefore, the wider the bandwidth, the greater the observed range resolution, whereas the higher the frequency of operation, the narrower the angle resolution. Increasing the frequency of the transmitted signal makes it much easier to achieve accurate angle and range resolution; however, it also has the added disadvantage of degrading the range or penetration depth. Usually, lower frequencies can penetrate deeper, but they provide very small angle and range resolution, due in part to the restrictions on the

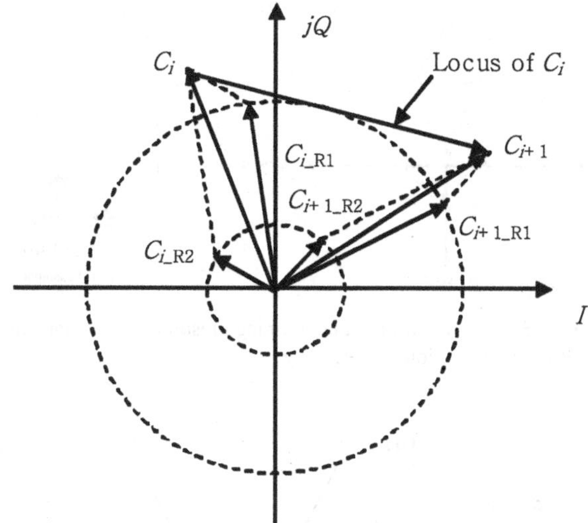

Fig. 3.4 Complex vectors $C_i = C_{i\_R1} + C_{i\_R2}$ for two stationary point targets moving along the locus

absolute bandwidth. Therefore, there is an inherent tradeoff involved in satisfying both the penetration depth and resolution requirements.

The frequency step is related to an ambiguous range, which is an unfolded range that can be defined by the sampling theory, while the pulse repetition interval affects the receiver's sensitivity.

It is apparent that the design parameters of SFCW radar sensors should be considered carefully and understood thoroughly in order to achieve an optimum design.

3.3.1 Angle and Range Resolution

The ability of a radar sensor to distinguish targets that are closely located depends on its resolution. As mentioned earlier, there are two kinds of resolution: angle and range resolution depending on the direction of observation from antenna as illustrated in Fig. 3.5. Angle and range resolution are therefore important in radar sensors' design and operation.

3.3.1.1 Angle Resolution

Angle resolution, also known as cross-range, horizontal, lateral or azimuth resolution, indicates the minimum angle two targets at the same range must be spaced apart in order to be distinguished. As an example, a radar sensor with an angle resolution of 5° can distinguish targets at the same range whose separation is larger than 5°. Angle resolution, as can be expected, is determined by the beam-width of

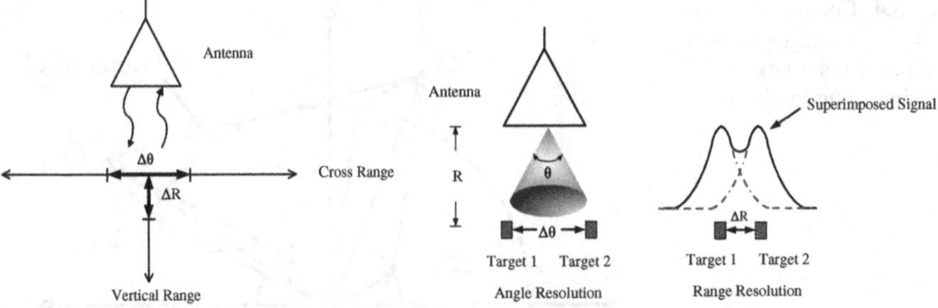

Fig. 3.5 Illustration of angle and range resolution of radar sensors. $\Delta\theta$ and ΔR denote the angle and range resolution, respectively

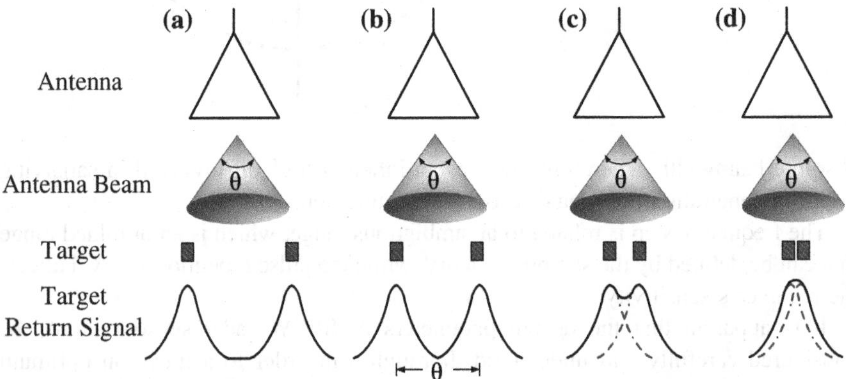

Fig. 3.6 Target distinguishing based on angle resolution: **a** targets easily resolved when (angle) separation >θ, **b** targets resolved when (angle) separation equal to θ, **c** targets difficultly resolved when (angle) separation <θ, and **d** targets unresolved when (angle) separation ≪θ. θ is the antenna's beam-width

the antenna and improves as the antenna's beam-width gets narrower. Figure 3.6 illustrates the angle-resolution phenomenon.

In the antenna's far-field region, occurring when the range between the antenna and targets is sufficiently large with respect to the operating wavelength, we can approximate the angle resolution as

$$\Delta R = R\theta \tag{3.22}$$

where θ is the antenna's beam-width (in radian) and R is the target's range in the far-field region. For aperture antennas, R can be approximated as

$$R \geq \frac{2D^2}{\lambda} \tag{3.23}$$

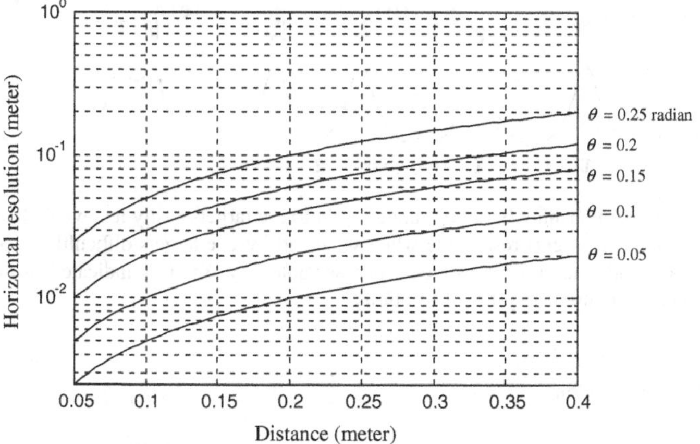

Fig. 3.7 Angle (horizontal) resolution versus range

where λ is the operating wavelength and D is the maximum dimension of the antenna. The beam-widths of antennas depend on the ratio of the antenna dimensions to the operating wavelength. For a given antenna size, the antenna's beam-width is proportional to the wavelength or inversely proportional to the frequency. High-frequency system therefore can focus energy into sharp beams, resulting in fine resolution and accurate determination of targets. Angle resolution improves as the antenna gets larger electrically.

Figure 3.7 shows the calculated angle resolution versus the (far-field) range for varying beam-width. In order to achieve angle resolution in cm, frequencies in Ka-band (26.5–40 GHz) need to be used. For instance, using a standard Ka-band waveguide horn antenna having about 0.26-radian 3-dB beam-width, we can estimate the angle resolution as 0.031 m for a target located at 0.12 m from a radar sensor.

3.3.1.2 Range Resolution

Range resolution determines how close two targets in different ranges can be spaced in order to be distinguished. For instance, a system having a range resolution of 10 cm can only distinguish targets separated by at least 10 cm in range. Figure 3.8 illustrates the range-resolution phenomenon.

Consider pulse signals returned from targets (or synthetic pulses formed from targets' returned sinusoidal signals in SFCW radar sensors), the narrow these pulses, the better targets can be resolved. For the targets to be resolved, they must be separated in time by an amount equal to the pulse duration τ of the received pulse as shown in Fig. 3.8 and elaborated in Fig. 3.9. This is equivalent to a range difference of

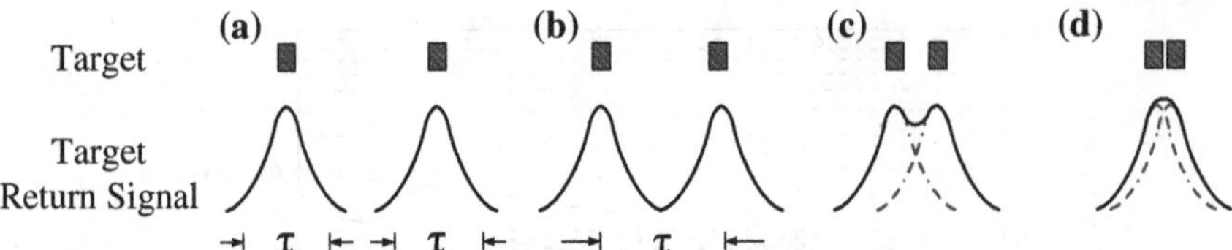

Fig. 3.8 Target distinguishing based on range resolution: **a** targets easily resolved when separated by >τ (pulse width), **b** targets resolvable when separated by τ, **c** targets difficultly resolved when separated by <τ, and **d** targets unresolved when separated by ≪τ. τ as indicated in the figure is loosely defined—it is used only for illustration

Fig. 3.9 Range resolution

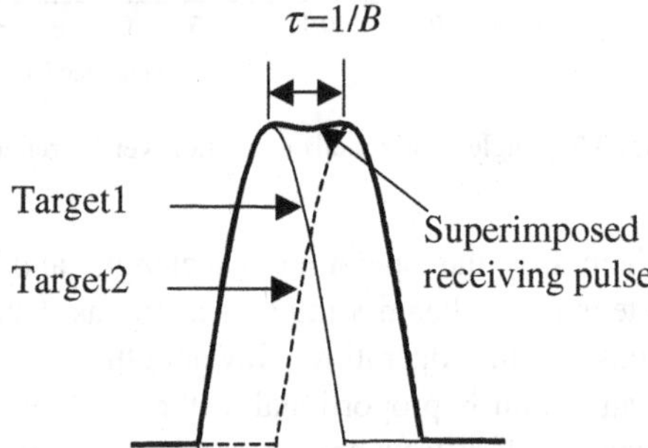

$$\Delta R = \frac{v\tau}{2} \tag{3.24}$$

which dictates the range resolution. The (absolute) bandwidth of a received pulse with a 4-dB pulse width of τ can be approximated as

$$B \cong \frac{1}{\tau} \tag{3.25}$$

Substituting (3.25) into (3.24) gives the range resolution

$$\Delta R \cong \frac{v}{2B} \tag{3.26}$$

Figure 3.10 illustrates the synthetic pulses caused by two targets at $R1$ and $R2$ that are superimposed. For simplicity without loss of generality, we assume these synthetic pulses have the same magnitude. The range resolution ΔR for the target responses seen in Fig. 3.10, which distinguishes two targets in range, can be defined as the range difference $R2 - R1$. The main lobe's null occurs when $n = n_{p1} + M/N$, resulting in the cell number $n_{p2} = n_{p1} + M/N$, where n_{p2} corresponding to the

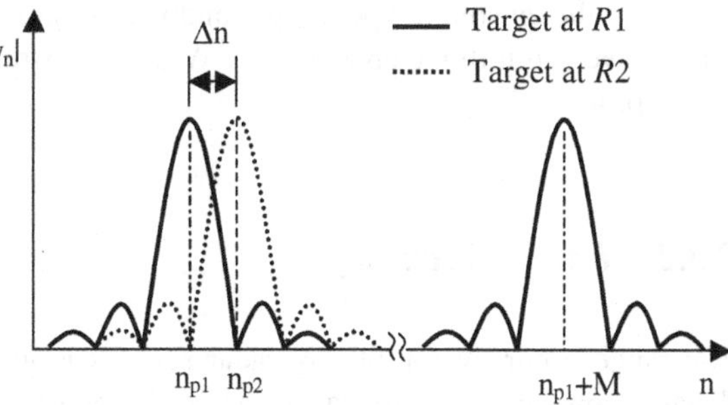

Fig. 3.10 Range resolution as defined by the main lobe's null

main-lobe's peak of the target at $R2$ coincides with the main lobe's null of the target at $R1$. Consequently, the range resolution ΔR is given by

$$\Delta R = R2 - R1 = \frac{\left(n_{p1} + \frac{M}{N}\right)v}{2M\Delta f} - \frac{n_{p1}v}{2M\Delta f} = \frac{v}{2N\Delta f} \tag{3.27}$$

which is identical to the range resolution given in (1.1) and (3.26) with the operating bandwidth $B = N\Delta f$.

Figure 3.11 shows the calculated range resolution as a function of the bandwidth for different relative dielectric constants (ε_r) of the transmission medium when a Hamming window factor (=1.33) is applied. According to these simulation results, the required bandwidth should be at least 4 GHz to achieve the vertical resolution in inches for subsurface sensing of pavements whose materials are shown in Table 2.1. However, with the theoretical range resolution, it is difficult to distinguish two synthetic pulses clearly, especially when these are superimposed. To overcome this

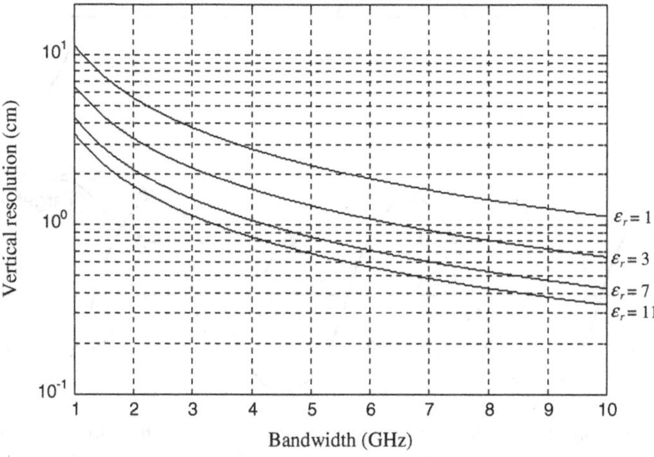

Fig. 3.11 Range resolution versus bandwidth $B = N\Delta f$

problem, the absolute operating bandwidths should be 5 and 8 GHz for the microwave and millimeter-wave SFCW radar sensors, respectively, to be described in Chap. 4.

3.3.2 Range Accuracy

The range accuracy for SFCW radar sensors is given in (3.17). Range accuracy dictates how accurate one can measure the range and is different from range resolution. The *rms* range error can be approximately derived for radar sensors as

$$\delta R = \frac{\Delta R}{\sqrt{2\left(\frac{S}{N}\right)}} = \frac{v}{2B\sqrt{2\left(\frac{S}{N}\right)}} \tag{3.28}$$

As can be seen, the range accuracy depends on the RF bandwidth of radar sensors; as the bandwidth is increased, the range error is reduced. Furthermore, the range error is inversely proportional to $\sqrt{S/N}$, indicating that more noise produces less accuracy in range, which is expected, as the higher the noise, the less perfect the pulse or waveform shape is. Figure 3.12 illustrates the dependence of the range error δR on bandwidth. As compared to the return signal corresponding to a single frequency, the double-frequency return signals allow a target to be more accurately

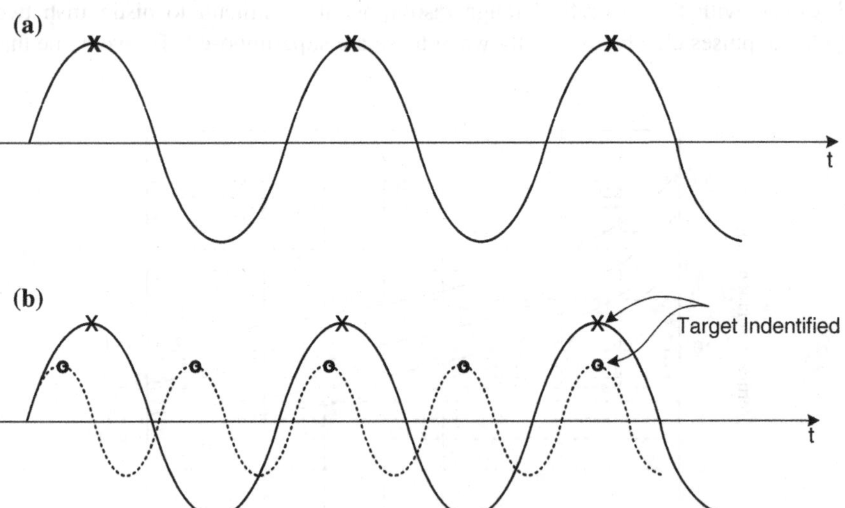

Fig. 3.12 Illustration of dependence of range accuracy on bandwidth with return signals for a single transmitting frequency (**a**) and for two transmitting frequencies f_1 and f_2 (**b**). The "×" and "o" indicate possible target locations corresponding to a phase measurement of 90°. When the "×" and "o" appear close to each other, a corresponding target can be identified

determined. As can be induced, when multiple frequencies are transmitted, or the operating bandwidth is large, the accuracy and un-ambiguity of the position of target is improved.

The angle accuracy or cross-range accuracy is different from the angle resolution and can be derived approximately as

$$\delta R = \frac{\Delta\theta}{3\sqrt{\frac{S}{N}}} = \frac{R\theta}{3\sqrt{\frac{S}{N}}} \tag{3.29}$$

where θ is the antenna's 3-dB beam-width.

3.3.3 Ambiguous Range

As seen from (3.19), the phase ϕ_i of the complex *I*/Q vectors depends on the frequency f_i. The resulting phases of the complex *I*/Q vectors are in the range of 2π. We consider two targets located at R_1 and R_2 that produce phases ϕ_{i_R1} and ϕ_{i_R2} at f_i, respectively. The phase differences associated with R_1 and R_2 between two successive frequencies f_i and f_{i+1} are $\Delta\phi_{R1} = \phi_{i\_R1} - \phi_{i+1\_R1}$ and $\Delta\phi_{R2} = \phi_{i\_R2} - \phi_{i+1\_R2}$, respectively. If these phases are equal, then the two targets tend to appear at the same location, causing ambiguity.

The phase differences $\Delta\phi_{R1}$ and $\Delta\phi_{R2}$ can also be obtained from (3.18) as

$$\Delta\phi_{R1} = -\frac{4\pi R_1 \Delta f}{v} \tag{3.30a}$$

$$\Delta\phi_{R2} = -\frac{4\pi R_2 \Delta f}{v} \tag{3.30b}$$

If $\Delta\phi_{R1} = \Delta\phi_{R2} \pm 2\pi n$, $n = 1, 2, 3, \ldots$, then the two targets are ambiguous. From (3.30a) and (3.30b), the ambiguous range R_u of SFCW radar sensors can be found as [23]

$$R_u = |R_1 - R_2| = \frac{v}{2\Delta f} \tag{3.31}$$

which shows that the ambiguous range is determined by the frequency step Δf of SFCW radar sensors.

The ambiguous range can also be found by using the sampling theory [37]. If a signal with bandwidth B is sampled with a sampling time Δt, the signal is replicated every $n(1/\Delta t)$ Hz in the frequency domain, where n is an integer, as shown in Fig. 3.13a, b. In order to avoid aliasing, the bandwidth B must be less than one-half of the inverse of the sampling time (i.e., $B \leq 1/2\Delta t$). Similarly, using the duality of the sampling theory, it can be inferred that the range R of SFCW radar sensors must

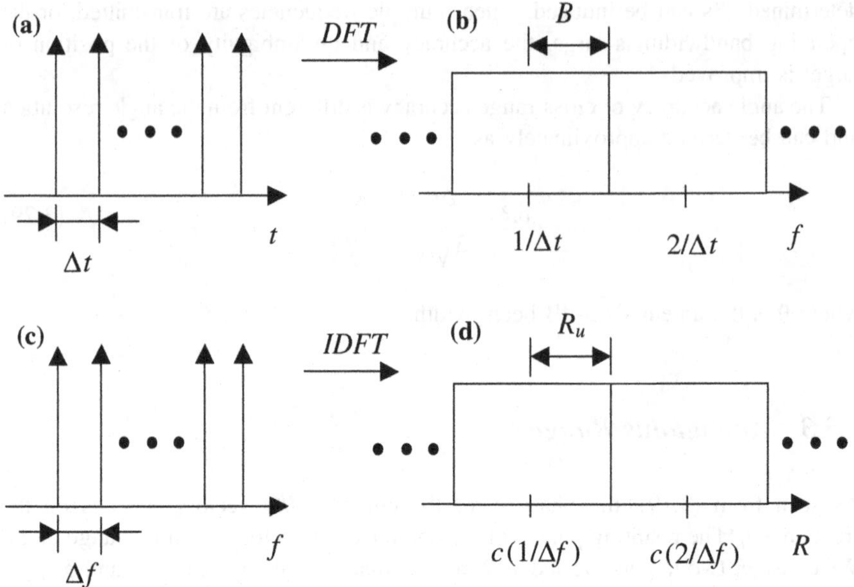

Fig. 3.13 Nyquist sampling to avoid aliasing: **a** time domain samples, **b** frequency domain of (**a**) through the DFT, **c** SFCW radar sensor's signals in the frequency domain, and **d** the range domain of (**c**) through the IDFT

be less than one-half of an inverse of the frequency step times the speed of the signal (i.e., $R \leq v/2\Delta f$) as shown in Fig. 3.13c, d. Thus, the resulting ambiguous range R_u of SFCW radar sensors is the same as given by (3.31).

Figure 3.14 shows the ambiguous ranges versus the frequency steps for various relative dielectric constants of the propagating medium. These simulation results show that the narrower the frequency step, the greater the ambiguous range as reflected in (3.31).

For the microwave and millimeter-wave SFCW radar sensors to be described in Chap. 4, a frequency step of 10 MHz is used for a reasonable ambiguous range, in spite of having an added disadvantage of requiring a long sweeping time for the signal source to cover the entire bandwidth. The rationale for this choice is further supported by the fact that the frequency synthesizer available for these SFCW radar sensors doesn't allow for the generation of arbitrary frequency steps in the vicinity of 10 MHz. Although other frequency steps such as 1 and 100 MHz can also be used with the available frequency synthesizer, they have significant limitations. The 1-MHz frequency step is deemed too narrow, and hence taking too long to sweep the entire operating frequency ranges, while the 100-MHz step is so large that the ambiguous range becomes so small of only 1.5 m.

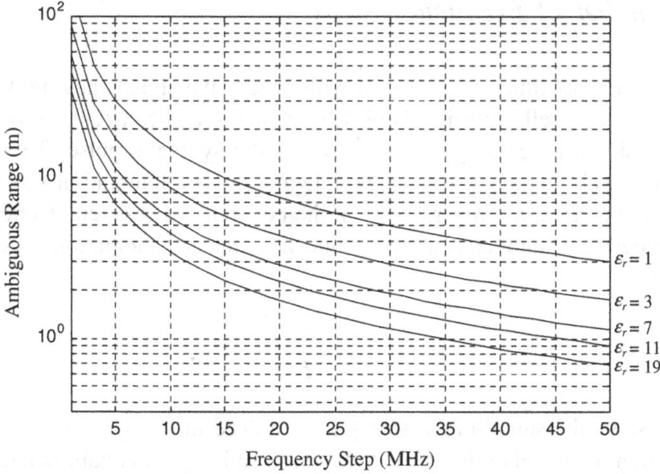

Fig. 3.14 Ambiguous range versus frequency step of SFCW radar sensors for different relative dielectric constants of the propagating medium

3.3.4 Pulse Repetition Interval

SFCW radar sensors need to transmit a complete set of stepped-frequency signals and capture the corresponding return signals to find the range information of targets. For each signal transmitted with a single frequency, the corresponding reflected signal is received during a particular pulse repetition interval (*PRI*). For coherent demodulation, the *PRI* should be at least larger than the two-way travel time to the target. Therefore, the *PRI* corresponding to a single stepped-frequency must be larger than the two-way travel time to the furthest target at *R*, which can be estimated as

$$PRI \geq \frac{2R}{v} \tag{3.32}$$

However, as the *PRI* should be considered up to the ambiguous range, it can be inferred from (3.31) and (3.32) that the *PRI* related to the frequency step is given by [23]

$$PRI \geq \frac{1}{\Delta f} \tag{3.33}$$

From (3.33), the minimum required *PRI* should be greater than 0.1 μs if the frequency step is set to 10 MHz. Thus, if a *PRI* of 50 μs is used with a frequency step of 10 MHz in a bandwidth of 5 GHz, the swept time of the entire bandwidth would be 25 ms.

3.3.5 Number of Frequency Steps

SFCW radar sensors illuminate a target with successive signals having N different frequencies, receive reflected signals of N frequencies, and coherently process these signals in a signal processing block to extract the synthetic pulse. Therefore, its process gain is said to be N if there is no integration loss. Integration loss is caused by a window function, an imperfection in the coherent process, etc. Generally, the effective integration number N_{eff} taking into account the integration loss is given by [38]

$$N_{eff} = \frac{N}{L_i} \tag{3.34}$$

which represents the signal-processing gain achieved in the system.

As an example, the Hamming window yields an integration gain (=$1/L_i$) of 0.54 [38]. For SFCW radar sensors operations involving multiple media, a complete coherent process is achieved when the dispersion effects of the propagating media are compensated for by signal processing with media's known properties.

3.4 System Performance Factor

The system performance factor as defined in (2.67) and derived in several equations in Chap. 2 is one of the most important parameters in the radar equation for estimating the range or penetration depth of radar sensors. Taking into account the effective integration number N_{eff} given in (3.34), we can modify the system performance factors derived in (2.75) and (2.79) of Chap. 2 for SFCW radar sensors as

$$SF = \frac{64\pi^2\left(\frac{R}{\cos\phi_{i1}} + \sum_{l=1}^{n-1}\frac{x_l}{\cos\phi_{tll-1}}\right)^2}{G^2\lambda^2 L' N_{eff}\Gamma^2_{nn-1}\left[\prod_{m=1}^{n-1}\mathrm{T}^2_{mm-1}\mathrm{T}^2_{m-1m}\exp\left(-\frac{4\alpha_m d_m}{\cos\phi_{tmm-1}}\right)\right]} \tag{3.35}$$

and

$$SF = \frac{64\pi^3\left(\frac{R}{\cos\phi_{i1}} + \frac{x_{1\max}}{\cos\phi_{t10}}\right)^4}{G^2\lambda^2\sigma N_{eff}\mathrm{T}^2_{10}\mathrm{T}^2_{01}L'\exp\left(-\frac{4\alpha_1 d_{1\max}}{\cos\phi_{t10}}\right)} \tag{3.36}$$

which can be used for evaluating the performance of SFCW radar sensors and estimate the maximum range for sensing multiple layers, such as pavements, and buried objects under a surface such as mines, respectively.

In practical radar sensors, the system performance factor can be limited by the actual receiver dynamic range, as discussed below. Hence, it is necessary to incorporate a correction into the system performance factor.

The maximum available dynamic range, *DRmax*, of a receiver is the ratio of the maximum available receiving power, *Pr,max*, that the receiver can tolerate without causing a distortion to the receiver's sensitivity which satisfies a specified S/N at the output of the receiver. The upper limit of the maximum available (compression-free) dynamic range is determined by the 1-dB compression point $P_{1\text{dB}}$ of the receiver front-end's low-noise amplifier (LNA) in order to avoid its saturation, while the lower limit is determined by the receiver's sensitivity. For safety considerations in practical systems, the maximum available receiver power needs to be below the 1-dB compression point of the LNA.

The maximum available receiving power of a system's receiver occurs when the system is directed toward a metal plate, which is typically done during a calibration process. If the transmission loss L_t is considered to be the difference between the transmitted and the received power when the corresponding antennae are directed onto a metal plate placed at a stand-off distance, *R*, as illustrated in Fig. 3.15, then it is found that the maximum available transmitting power $P_{t,max}$ can be estimated from the maximum available receiving power as

$$P_{t,\max} = P_{r,\max} + L_t \leq (P_{1\text{dB}} + L_t) \quad (\text{dB}) \tag{3.37}$$

It should be noted that the above analysis is valid only if the maximum receiving power is less than the saturating power of the receiver.

The transmission loss L_t (as noted by S_{21} in Fig. 3.15) is caused by the spreading loss, the antenna's mismatch and efficiency, and others practical losses arising from connectors and cables. The transmission loss can be approximately calculated using an accurate EM simulator or accurately measured using a network analyzer if antennae are available, as shown in Fig. 3.15.

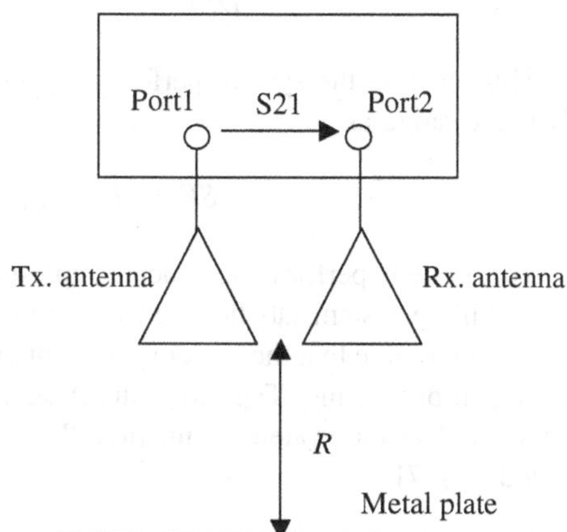

Fig. 3.15 Measurement of the transmission loss L_t using a network analyzer. *R* is the stand-off distance between the antennas and a metal plate

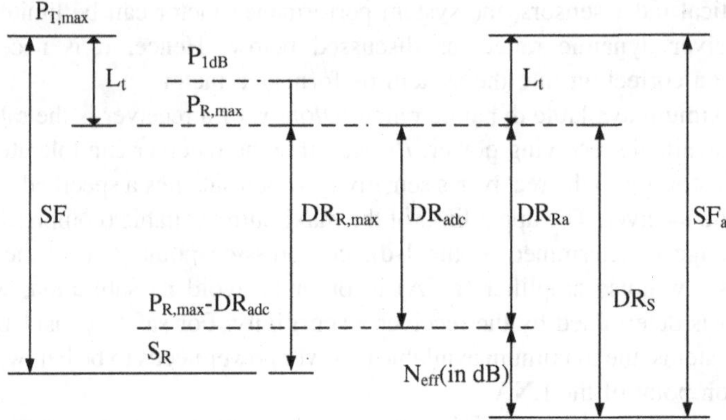

Fig. 3.16 Graphical analysis of the system performance factor and dynamic range when $DR_{adc} \leq DR_{R,max}$

The instantaneous bandwidth of the SFCW radar sensor is equal to the inverse of the *PRI*, as the frequency band of a single frequency *f* during time τ is equal to 1/τ around the center frequency *f* [37]. Thus, the instantaneous bandwidth of the input signal at the receiver is much less than the total bandwidth *B*, which results in a low sensitivity level at the receiver, as defined by Eq. (2.35).

Figure 3.16 illustrates the system performance factor and dynamic range of a system. The system performance factor (in dB) can be found using (2.67) and (3.37) as

$$SF = (P_{t,\max} - S_R) = (P_{r,\max} + L_t - S_R) \quad \text{(dB)} \tag{3.38}$$

The receiver's maximum available dynamic range, $DR_{R,max}$, can be defined as the difference between the maximum available receiving power and the receiver's sensitivity:

$$DR_{R,\max} = (P_{r,\max} - S_R) \quad \text{(dB)} \tag{3.39}$$

This leads to the system performance factor in term of the maximum available dynamic range as

$$SF = (DR_{R,\max} + L_t) \quad \text{(dB)} \tag{3.40}$$

The system performance factor represents the maximum performance of the system if the system satisfies the maximum available dynamic range. However, it is important to note that the system also contains analog-to-digital converters (ADC's) for signal processing. Therefore, the ADC's dynamic range, DR_{adc}, should also be considered in the system evaluation. The ADC's dynamic range can be approximated as [37]

$$DR_{adc} = 6N \quad \text{(dB)} \tag{3.41}$$

where N is the number of bits of the ADC. Therefore, the receiver's available dynamic range, DR_{Ra}, is limited by either the receiver's maximum available dynamic range or the ADC's dynamic range, whichever is narrower. A signal processing gain typically implemented in systems, however, increases the receiver's dynamic range. The system dynamic range, DR_S, can therefore be defined as

$$DR_S = \left[DR_{Ra} + 10\log\left(N_{eff}\right)\right] \quad \text{(dB)} \tag{3.42}$$

where N_{eff} is the signal-processing gain achieved in the system. As a result, the actual system performance factor can be obtained as

$$SF_a = (DR_S + L_t) \quad \text{(dB)} \tag{3.43}$$

The range of the system can be estimated more accurately using the radar equation incorporating the actual system performance factor obtained by (3.43).

3.5 Estimation of Range or Penetration Depth

3.5.1 Estimation of Penetration Depth in Multi-layer Targets

As an example to illustrate the estimation of the maximum range or penetration depth in the sensing of multi-layer targets, we consider a pavement structure consisting of asphalt, base and subgrade layers as shown in Fig. 3.17. Furthermore, we assume 3-GHz operation for the SFCW radar sensor, which will be presented in Chap. 4, and consider the effect of the signal's incident angle for the sensor in the calculations.

The parameters used for the calculations are listed in Table 3.1 where the system loss L is set to 19 dB, which includes the antenna efficiency (6 dB per a resistively loaded antenna and 12 dB for a pair of such antennas), the antenna mismatch (1 dB per antenna or 2 dB for a pair of antennas), and the other losses (5 dB) due to cables, connectors, etc. [1]. The signal-processing gain is 24 dB, while the Hamming window function and 500 frequency steps are used on the signal processing without the integration loss. The ADC in the data acquisition (DAQ) board of LabView[1] has a resolution of 12 bits per sample, which leads to a dynamic range of 72 dB. The measured transmission loss is 25 dB after incorporating the antennas (which will be discussed later in Chap. 5). The actual system performance factor is hence estimated to be 121 dB using (3.42).

[1]A graphical software system called "Virtual Instruments" from National Instruments.

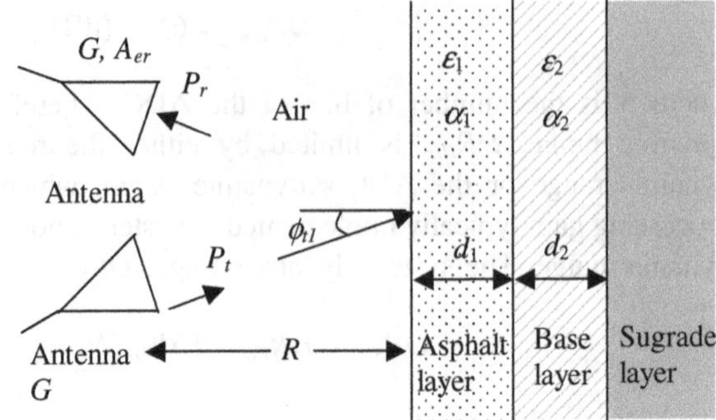

Fig. 3.17 Pavement layers used for estimating penetration depths

Table 3.1 Parameters used in the calculations to estimate the penetration depth in the pavement structure shown in Fig. 3.17

Electrical properties of pavement layers at 3 GHz		
Asphalt layer	ε'_{r1}	5–7
	ε''_{r1}	0.03–0.05
	α	0.05–0.5 (Np/m)
Base layer	ε'_{r1}	8–12
	ε''_{r1}	0.3–0.8
	α	3–9 (Np/m)
Subgrade layer	ε'_{r3}	20
Radar sensor parameters		
Antenna gain	G	10 dB
Wavelength at 3 GHz	λ	0.1 m
System loss	L	19 dB
Process gain	G_p	24 dB
Incident angle	ϕ_{t1}	20°
Stand-off distance	R	0.2 m

Figure 3.18 illustrates the maximum penetration depth (or maximum detectable thickness) of the asphalt layer versus the actual system performance factor with different attenuation constants. The results show that the attenuation constant and the actual system performance factor significantly affect the penetration depth. According to the simulation results, this SFCW radar sensor based on the actual system performance factor of 121 dB can detect the thickness of the asphalt layer in the range of 2.3–9.5 m depending on the attenuation constant.

The simulation results shown in Fig. 3.19 represent the maximum penetration depth (or maximum detectable thickness) of an asphalt layer versus the actual system performance factor with incident angles of 0 and 20°, where the attenuation constant of the asphalt layer is fixed at 0.3 (Np/m). The results show that the incident angle of 20° does not significantly affect the maximum detectable range as compared to the normal incidence.

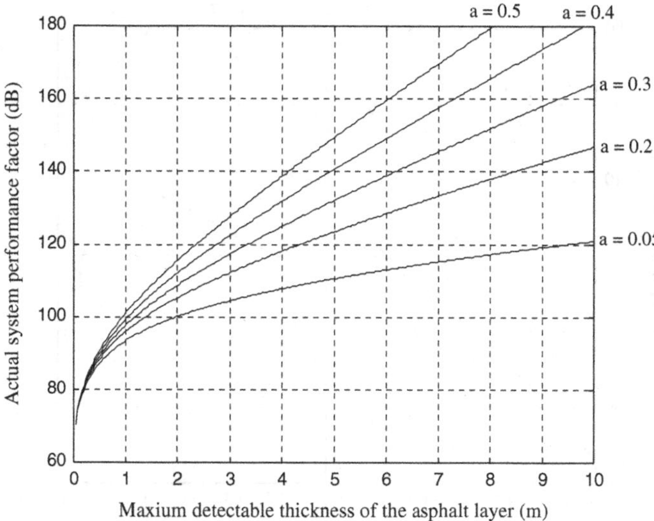

Fig. 3.18 Maximum penetration depth (or maximum detectable thickness) of the asphalt layer versus the actual system performance factor for different attenuation constants. The constant "a" denotes the attenuation constant (Np/m)

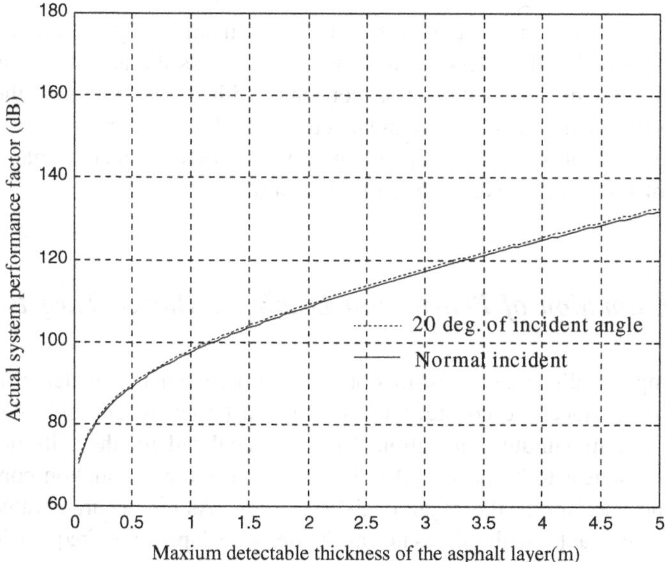

Fig. 3.19 Maximum penetration depth (or maximum detectable thickness) of the asphalt layer versus the actual system performance factor for 0 and 20° incident angles with the attenuation constant of the asphalt layer as 0.3 (Np/m)

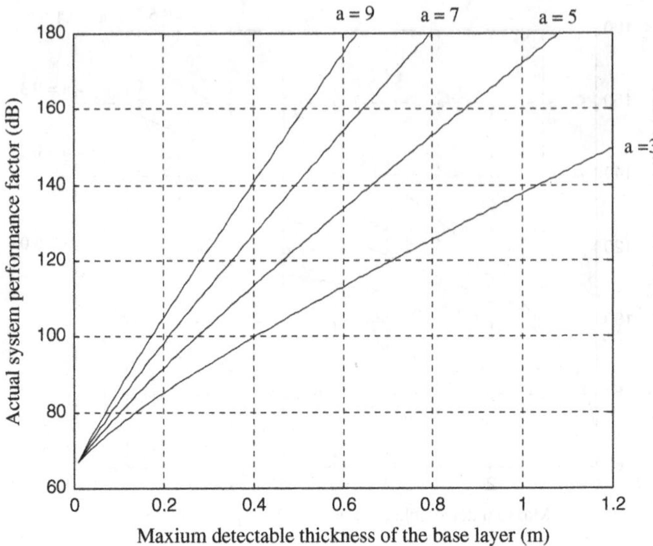

Fig. 3.20 Maximum penetration depth (or maximum detectable thickness) of the base layer versus the actual system performance factor with different attenuation constants, where "a" denotes the attenuation constant (Np/m) and the thickness of the asphalt layer is 3 inches

Figure 3.20 shows the simulation results for the maximum penetration depth (or maximum detectable thickness) of the base layer versus the actual system performance factor for different attenuation constants with the thickness of the asphalt layer fixed at 3 inches. This SFCW radar sensor can detect the thickness of the base layer in the range of 0.2–0.4 m, depending upon the attenuation constants, when the thickness of the asphalt layer is fixed to 3 inches.

3.5.2 Estimation of Penetration Depth for Buried Targets

As an example to illustrate the estimation of maximum penetration depth for targets underneath a surface, we consider a buried metal target under sand as shown in Fig. 3.21. The maximum penetration depth is calculated for the millimeter-wave SFCW radar sensor to be presented in Chap. 4 where the attenuation constants of the dry sand used are in the range of 3–70 (Np/m). As the accurate values of the attenuation constants of the dry sand at the sensor's operating frequencies in the Ka-band (26.5–40 GHz) are not available, we assume these are the same as those at 1 GHz (0.1–2.3) and 100 MHz (0.01–0.23) as mentioned in [1]. These assumptions are useful even they are not very accurate in Ka-band. The parameters used for the

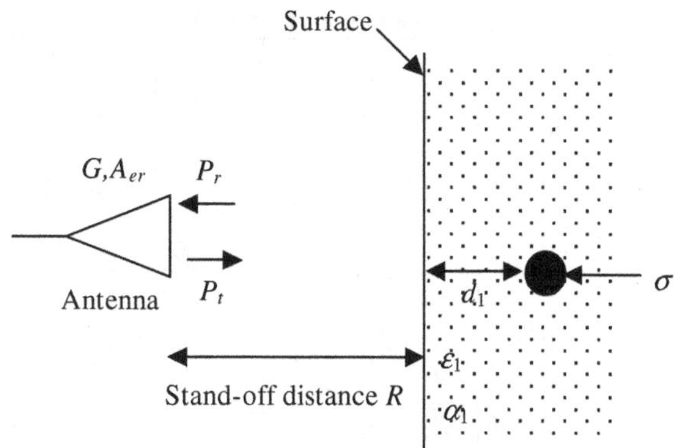

Fig. 3.21 Buried target used for estimating the penetration depth

Table 3.2 Parameters used for estimating the penetration depth of the millimeter-wave SFCW radar sensor for a buried object shown in Fig. 3.21

Electrical properties of the material used at 30 GHz		
Dry sand	ε'_{r1}	3–6
	α	3–70
RCS of the target	σ	0.0019 m^2
Radar sensor parameters		
Antenna gain	G	24 dB
Wave length	λ	0.01 m
System loss	L	17 dB
Process gain	G_p	23 dB
Stand-off distance	R	0.1 m

calculations are listed in Table 3.2, where the system loss is set to 17 dB, which includes the antenna efficiency of 4 dB and the antenna mismatch of 2 dB [1], and losses of connectors, cables, adapters and circulator of 9 dB. The process gain is 23 dB, while the Hamming window function and 400 frequency steps are applied to the signal processing without the integration loss. The measured transmission loss is 13 dB when a Ka-band waveguide horn antenna is used. The actual system performance factor is estimated to be 108 dB where the ADC's dynamic range is 72 dB.

Figure 3.22 shows the maximum detectable depth versus the actual system performance factor for a spherical object (radius = 0.025 m) buried under the sand with different attenuation constants. The results show that the millimeter-wave SFCW radar sensor can detect the buried spherical target in the range of 0.05–0.5 m, depending upon the attenuation constant of the sand.

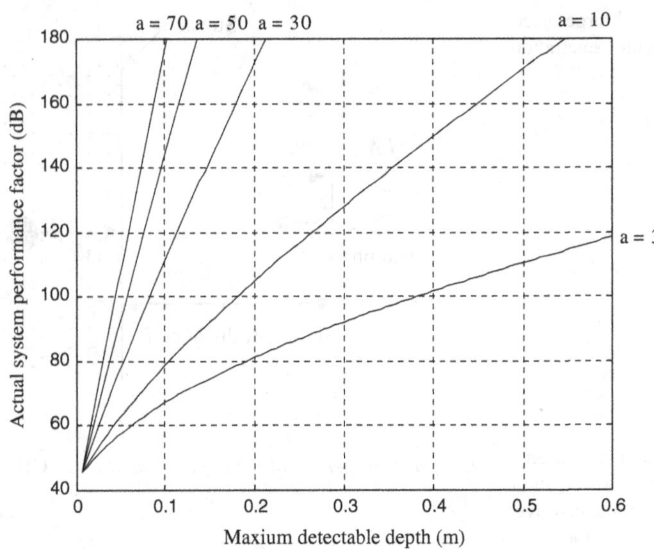

Fig. 3.22 Maximum detectable depth versus the actual system performance factor with different attenuation constants for a spherical object (radius = 0.025 m) buried under the sand (ε_{r1} = 3) where "a" denotes the attenuation constant (Np/m) of the sand

3.6 Summary

The analysis of SFCW radar sensors is presented in this chapter. Specifically, it addresses the system's important parameters including transmitted and received signals, down-converted I and Q signals, synthetic pulse, angle and range resolutions, frequency step, number of frequency steps, total bandwidth, range or penetration depth, range accuracy, range ambiguity, pulse repetition interval, dynamic range, and system performance factor. Maximum ranges or penetration depths in sensing multi-layer and buried targets are also discussed. The presented materials, although concise, should provide sufficient information for RF engineers to undertake the analysis of their own SFCW radar sensors and help design these sensors for sensing applications.

Chapter 4
Development of Stepped-Frequency Continuous-Wave Radar Sensors

4.1 Introduction

This chapter presents two different stepped-frequency continuous-wave (SFCW) radar sensors: a millimeter-wave SFCW radar sensor working across 29.72–37.7 GHz and a microwave SFCW radar sensor operating from 0.6 to 5.6 GHz. A millimeter-wave SFCW radar sensor can achieve both fine range and angle resolutions for surface and subsurface sensing. Its range or penetration distance, however, is rather limited due to small RF power typically available at millimeter-wave frequencies. On the other hand, a microwave SFCW radar sensor operating at lower frequencies can satisfy both deep penetration and fine range resolution simultaneously for subsurface sensing, and is thus attractive for subsurface evaluation.

Complete developments of these sensors are described in this chapter, including hardware (transceiver and antenna) design, signal processing, and integration. Both of them are realized using microwave integrated circuits (MICs) and microwave monolithic integrated circuits (MMICs) in single packages, and can be used for surface and subsurface sensing. The applications of the millimeter-wave SFCW radar sensor for surface profiling, monitoring liquid levels and detecting and localizing buried objects are described in Chap. 5. The use of the microwave SFCW radar sensor for characterizing pavement structures is covered in Chap. 5.

The chapter begins with discussions of the homodyne and super-heterodyne architectures and an analysis of the quadrature detectors, which play the important role of recovering RF signals for the systems. The designs of the millimeter-wave and microwave SFCW radar sensor transceivers are presented including their block diagrams, operations and components, and analyses of the receivers and transmitters. The design of the 0.5–10 GHz microwave and 26.5–40 GHz millimeter-wave microstrip quasi-horn antennas for the sensors are then addressed. Finally, the development of the sensors' signal processing is presented, which includes data acquisition, synchronization, compensation for the amplitude and phase errors due

C. Nguyen and J. Park, *Stepped-Frequency Radar Sensors*,
SpringerBriefs in Electrical and Computer Engineering,
DOI 10.1007/978-3-319-12271-7_4

to inherent imperfections of the systems, and generation of the synthetic pulses of targets for extracting target information.

4.2 SFCW Radar Sensors

Figures 4.1a, b shows the block diagrams of the SFCW radar sensors based on the homodyne and the super-heterodyne architectures, respectively. The major differences between these two systems, as can be seen, are the architectures and functions of the transceivers. The homodyne system down-converts the RF input signal once, essentially executing a direct conversion, in the transceiver while the super-heterodyne system down-converts the input signal twice to produce a base-band signal. All are done coherently.

The homodyne system, also called a zero *IF* system, down-converts the input signal directly into base-band in-phase (*I*) and quadrature (*Q*) signals using a quadrature detector or mixer. Thus, a homodyne system with a wide RF bandwidth requires a wideband quadrature detector operating at high frequencies. This quadrature detector could have a large varying response, especially the phase, over a wide RF bandwidth for its constituent 90° phase shifter, which would lead to large

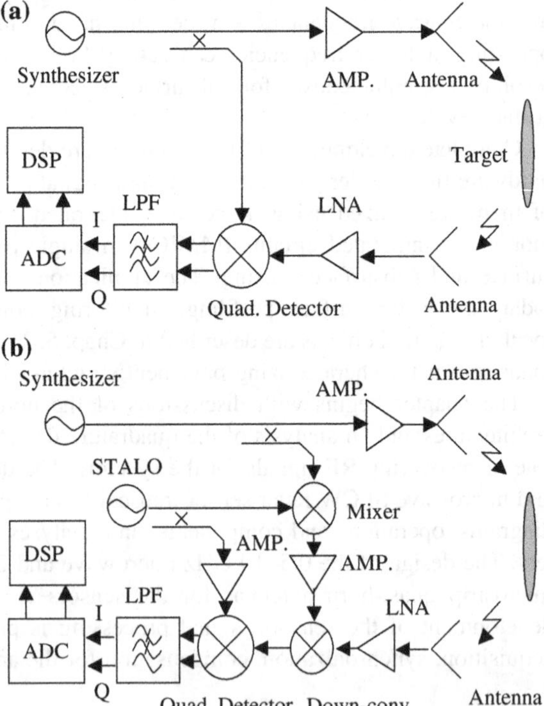

Fig. 4.1 System block diagrams of SFCW radar sensors: **a** homodyne and **b** super-heterodyne architectures

I/Q imbalances over the RF frequency band. On the other hand, in the super-heterodyne SFCW radar sensors, the input signal is first down-converted to an intermediate-frequency (*IF*) signal, which is then further down-converted into base-band *I* and *Q* signals via a quadrature detector. The IF down-converted frequency-band typically has a single frequency or narrow bandwidth, making it possible to use a narrow-band quadrature detector operating at low frequencies. Such quadrature detector has a relatively constant response for the 90° phase shifter, thereby leading to less *I/Q* imbalances and hence easy compensation or correction for the I/Q errors. Consequently, the super-heterodyne SFCW radar sensor is preferred over the homodyne counterpart for wide RF frequency bands, especially at millimeter-wave frequencies, even it is more complex. The microwave and millimeter-wave SFCW radar sensors described in this chapter implement the super-heterodyne architecture.

4.2.1 *Quadrature Detectors*

The most critical function of a RF receiver, whether superheterodyne or homodyne, is to fully recover a RF signal. The recovery may be done through quadrature detection accomplished by a quadrature detector as mentioned above. Quadrature detector is basically used to measure both the amplitude and phase of a received signal relative to a transmitted signal by determining the two orthogonal components, namely the *I* and *Q* components, of the down-converted signal.

4.2.1.1 Principle of Quadrature Detectors

Block diagrams of quadrature detectors are shown in Fig. 4.2, with each consisting of two identical mixers (Mixer I and Mixer Q) of any type in the I and Q channels. A quadrature detector requires a 90° phase shift in either the LO or RF signal path. However, practical quadrature detector structures usually employ a 90° phase shift in the LO path (Fig. 4.2a) due to a narrow LO bandwidth typically encountered in mixers which facilitates the design of the phase-shifting network. To illustrate the detector's operation, we consider the detector shown in Fig. 4.2a and assume it is an ideal component with all ideal constituents, hence perfect balance between the I and Q channels, and no leakage between the LO and RF ports of the constituent mixers. The LO signal splits into two signals of equal amplitude and 90° phase difference. The 90° out-of-phase is typically achieved using a 90° 3-dB hybrid. The LO signal driving Mixer I can be expressed as

$$v_{LO}^{I}(t) = V_{LO}\cos(2\pi ft) \tag{4.1}$$

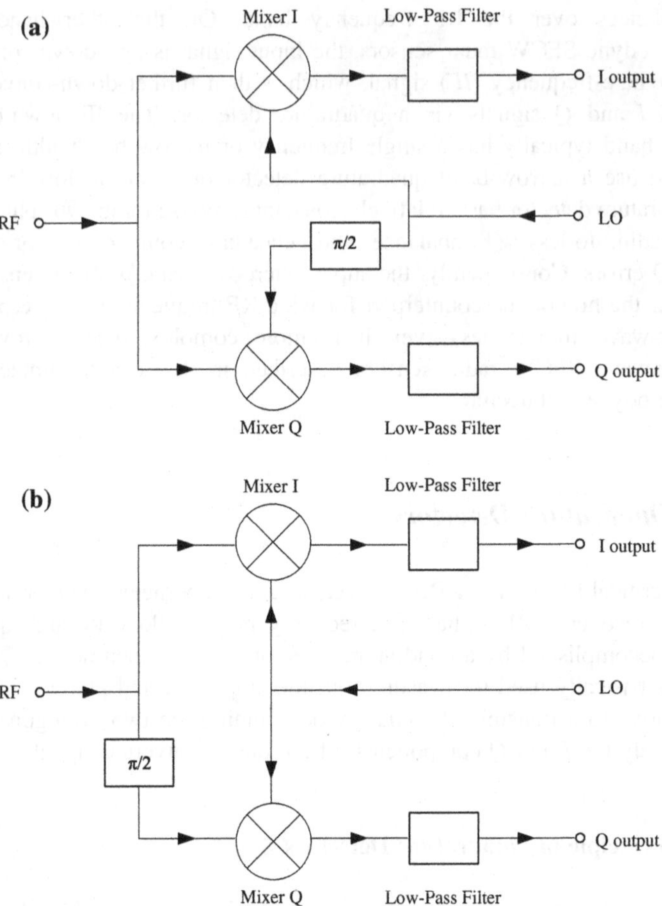

Fig. 4.2 Block diagrams of quadrature detectors with 90° phase shift in the LO path and equal phase in the RF path (**a**) and 90° phase shift in the RF path and equal phase in the LO (**b**)

where f is the frequency of the signal and V_{LO} is $1/\sqrt{2}$ of the amplitude of the LO signal pumping the quadrature detector (i.e., at the input of the 90° phase shifter), while that arriving at Mixer Q is

$$v_{LO}^{Q}(t) = V_{LO}\cos\left(2\pi ft - \frac{\pi}{2}\right) = V_{LO}\sin(2\pi ft) \tag{4.2}$$

which is delayed by 90° with respect to the LO signal at Mixer I. The RF signal splits into two signals of equal amplitude and phase using an in-phase splitter such as a power divider. The RF signals arriving at Mixer I and Mixer Q can be described as

$$v_{RF}^{I}(t) = v_{RF}^{Q}(t) = V_{RF}\cos[2\pi ft - \phi(t)] \tag{4.3}$$

where V_{RF} is $1/\sqrt{2}$ of the amplitude of the RF signal reaching the quadrature detector and $\phi(t) = 2\pi ft_R - \phi_o$ is the phase of the return signal at the mixer, assuming the received signal is from a target, with t_R representing the roundtrip time delay in signal propagation to and from the target and ϕ_o being the initial phase. $\phi(t)$ excludes the initial phase produced by the difference in the electrical lengths between the LO and RF paths leading to the quadrature detector. The contribution from the initial phase to the detected phase of the received signal can be eliminated by inserting a variable phase shifter in the LO or RF path to the mixer to nullify the initial phase.

The RF and LO signals mix at Mixer I to produce the I signal in the I channel:

$$\begin{aligned} v_I(t) &= V_{RF}\cos[2\pi ft - \phi(t)] \cdot V_{LO}\cos(2\pi ft) \\ &= \frac{1}{2}V_{RF}V_{LO}[\cos(4\pi ft - \phi) + \cos(\phi)] \end{aligned} \tag{4.4}$$

The second term in (4.4) represents the average of DC value and is proportional to the magnitudes of the LO and RF signals and *cosine* of the return signal's phase angle. The first term represents the second harmonic of the signal which can be removed by a low-pass filter. The I signal can be rewritten, after passing through a low-pass filter, as

$$v_I(t) = \frac{1}{2}V_{RF}V_{LO}\cos[\phi(t)] \tag{4.5}$$

Similarly, from the Q-channel, we get the Q signal after its second harmonic is filtered out:

$$v_Q(t) = \frac{1}{2}V_{RF}V_{LO}\sin[\phi(t)] \tag{4.6}$$

It can be seen that the I and Q signals are orthogonal to each other with an exact 90° phase difference between them. It is also recognized that both the final I and Q output signals are baseband signals. The I and Q signals completely represent the received RF signal $v_{RF}(t)$ and hence can be used to reconstruct it. We consider a (complex) exponential represented by

$$v(t) = v_I(t) + jv_Q(t) = \frac{1}{2}V_{RF}V_{LO}\{\cos[\phi(t)] + j\sin[\phi(t)]\} \tag{4.7}$$

It is recognized, from (4.5) and (4.6), that

$$V = \frac{1}{2}V_{RF}V_{LO} \tag{4.8}$$

and

$$\phi(t) = \tan^{-1}\left[\frac{v_Q(t)}{v_I(t)}\right] \tag{4.9}$$

are the amplitude and phase of the received signal, respectively. Hence, $v(t)$ in (4.7) indeed represents the (composite) received signal. We can rewrite this signal as

$$v(t) = v_I(t) + jv_Q(t) = Ve^{j\phi(t)} \tag{4.10}$$

from which V can be determined as

$$V = \sqrt{v_I^2(t) + v_Q^2(t)} \tag{4.11}$$

that, together with (4.9), show that the received signal can be reconstructed from the measured I and Q signals. The response of the output signal $v(t)$ is rotated in a circle on a complex plane. When the phase of the received signal is constant, i.e., having only a single value, (e.g., signal return from a single target in radar), $v(t)$ has only one value. In general, the phase of the received signal varies (e.g., signal return from multiple targets in radar sensors), causing the output signal to rotate in the counter-clockwise (CCW) or clockwise (CW) direction depending on whether the phase $\phi(t)$ is positive or negative, respectively.

4.2.1.2 Practical Quadrature Detectors

In practice, the in-phase splitter, such as a power divider, used for the RF signal does not provide equal magnitude and phase for the splitting signals, and the 90° phase shifter, such as a 90° hybrid, does not divide the LO signal exactly equal in amplitude and 90° out of phase. These imperfect components along with the difference between other (same) components in the I and Q channels, such as the constituent mixers, low-pass filters, transmission lines, lumped elements, cause mismatch between these channels. This mismatch results in amplitude and phase imbalances between the channels, which vary with frequency due to the frequency-dependence of all the components. This is the most fundamental and severe problem in quadrature detectors, widely known as the I/Q error, which limits the accuracy of measurement, particularly over a wide frequency range. The (frequency-dependent) I/Q error causes actual quadrature detectors to deviate from the ideal behavior and results in non-linear response. The nonlinear phase response of quadrature detectors is a critical problem in systems as it affects significantly the measurement accuracy. The effect is less in a super-heterodyne system as compared to a homodyne system since a single constant intermediate frequency is typically used for the quadrature detector, leading to a constant I/Q error over the operating frequency range. The instability of the frequency source also affects the measurement. This, however, should produce a negligible effect provided that the time delay between the transmit and receive signals of the system is short. Additionally, due to finite isolation between the ports of the constituent mixers, signal leakages between the LO and RF ports occur. The leakage of the RF signal to the LO port is typically negligible due to the small RF power as compared to that of the LO signal. However, the leakage of the LO signal onto the RF port can be significant. This LO-leaking signal mixes with the original LO signal to produce an additional signal, which results in a DC component in each of the output signals, known as the DC offset, worsening the nonlinearity caused by the amplitude and phase imbalances. These DC offset voltages, however, can be filtered out by using a band-pass filter. Inevitable errors are thus generated in practical (non-ideal) quadrature detectors, which can be severe at high RF frequencies such as those in the millimeter-wave range.

For simplicity without loss of generality, we assume the in-phase splitter used for the RF signal is perfect and all components in the I and Q channels are identical, so the mismatch is only caused by the 90° phase shifter. We also assume that effect of the phase noise of the LO signal source pumping the mixer is negligible; for short delay time between the transmit and receive signals, this assumption is actually valid. We can describe the LO signal arriving at Mixer I as

$$v_{LO}^{I}(t) = V_{LO}(1 + \delta V_{LO})\cos(2\pi f t) \tag{4.12}$$

or

$$v_{LO}^{I}(t) = (V_{LO} + \Delta V_{LO})\cos(2\pi f t) \tag{4.13}$$

where δV_{LO} represents the relative amplitude imbalance or relative loss (gain) imbalance between the I and Q channels and ΔV_{LO} denotes the absolute amplitude imbalance between the two channels at frequency f. They are related by

$$\delta V_{LO} = \frac{\Delta V_{LO}}{V_{LO}} \tag{4.14}$$

The LO signal at Mixer Q is given as

$$v_{LO}^{Q}(t) = V_{LO}\sin(2\pi f t + \Delta\phi) \tag{4.15}$$

where $\Delta\phi$ represents the (absolute) phase imbalance between the I and Q channels. We now consider two cases: with and without DC offsets in the channels.

No DC Offsets

The RF signals arriving at the constituent mixers are given in (4.3). The I signal produced by the I channel is obtained as

$$v_I(t) = \frac{1}{2}V_{RF}(V_{LO} + \Delta V_{LO})[\cos(4\pi ft - \phi) + \cos(\phi)] \tag{4.16}$$

which becomes, after the second harmonic is suppressed,

$$v_I(t) = \frac{1}{2}V_{RF}(V_{LO} + \Delta V_{LO})\cos[\phi(t)] \tag{4.17}$$

Similarly, we can write the Q signal emerging from the Q channel as

$$v_Q(t) = \frac{1}{2}V_{RF}V_{LO}[\sin(4\pi ft - \phi + \Delta\phi) + \sin(\phi + \Delta\phi)] \tag{4.18}$$

or, after the low-pass filter,

$$v_Q(t) = \frac{1}{2}V_{RF}V_{LO}\sin[\phi(t) + \Delta\phi] \tag{4.19}$$

Equations (4.17) and (4.19) show that, with the amplitude and phase imbalances, the I and Q signals are no longer orthogonal and balanced; i.e., they have phase difference deviating from 90° and unequal amplitude.

With DC Offsets

As mentioned earlier, the LO leakage from the LO port to the RF port of each constituent mixer causes DC offset. Under this condition, the RF signals appearing at the RF ports of Mixer I and Mixer Q in the I and Q channels, respectively, would consist of the original RF signal and the LO leakage and can hence be expressed as

$$v_{RF}^I(t) = V_{RF}\cos[2\pi ft - \phi(t)] + \alpha(V_{LO} + \Delta V_{LO})\cos(2\pi f_o t) \tag{4.20}$$

and

$$v_{RF}^Q(t) = V_{RF}\cos[2\pi ft - \phi(t)] + \beta V_{LO}\sin(2\pi f_o t + \Delta\phi) \tag{4.21}$$

where $0 < \alpha < 1$ and $0 < \beta < 1$.

Taking the product of v_{LO}^I and v_{RF}^I from (4.13) and (4.20), respectively, gives the I output signal:

$$\begin{aligned} v_I(t) = \frac{(V_{LO} + \Delta V_{LO})}{2}\{&V_{RF}[\cos(4\pi ft - \phi) + \cos\phi] \\ &+ \alpha(V_{LO} + \Delta V_{LO})[\cos(4\pi ft) + 1]\} \end{aligned} \tag{4.22}$$

which becomes, after its second harmonic is filtered out,

$$v_I(t) = \frac{(V_{LO} + \Delta V_{LO})}{2}[V_{RF}\cos\phi + \alpha(V_{LO} + \Delta V_{LO})] \tag{4.23}$$

The second term

$$V_{OSI} = \frac{\alpha(V_{LO} + \Delta V_{LO})^2}{2} \tag{4.24}$$

is the DC offset voltage of the I signal caused by the LO-to-RF leakage in Mixer I.
Similarly, the Q output signal can be determined from (4.15) and (4.21) as

$$\begin{aligned} v_Q(t) = \frac{V_{LO}}{2}\{&V_{RF}[\sin(4\pi ft - \phi + \Delta\phi) + \sin(\Delta\phi + \phi)] \\ &+ \beta V_{LO}[1 - \cos(4\pi ft + 2\Delta\phi)]\} \end{aligned} \tag{4.25}$$

or, after filtering out the second harmonic,

$$v_Q(t) = \frac{V_{LO}}{2}\{V_{RF}\sin(\phi + \Delta\phi) + \beta V_{LO}\} \tag{4.26}$$

where the second term

$$V_{OSQ}(t) = \frac{\beta V_{LO}^2}{2} \tag{4.27}$$

represents the DC offset of the Q signal caused by the LO-to-RF leakage in Mixer Q.

The I and Q output signals of a practical quadrature mixer can now be rewritten from (4.23)–(4.24) and (4.26)–(4.27), respectively, as

$$v_I(t) = (V + \Delta V)\cos(\phi) + V_{OSI} \tag{4.28}$$

and

$$v_Q(t) = V\sin(\phi + \Delta\phi) + V_{OSQ} \tag{4.29}$$

where $V \equiv V_{RF}V_{LO}/2$ and $\Delta V \equiv V_{RF}\Delta V_{LO}/2$.

The function of a quadrature detector, whether it is used in a homodyne or super-heterodyne system, is to directly down-convert a signal; i.e., it functions as a homodyne system itself. In addition to the amplitude- and phase-imbalance and DC offset issues seen in (4.28) and (4.29), the *1/f* noise contribution is also a critical problem in the direct down conversion. A simple way to overcome this problem is to slightly shift the LO frequency so that the frequency of the detector's output signals is located sufficiently far away from the *1/f* noise spectrum.

The phase of the received signal for practical quadrature detectors, considering the nonlinear phase response due to their phase and amplitude imbalances as well as the DC offset voltages, can be obtained by solving Eqs. (4.28)–(4.29) for $\phi(t)$ as

$$\phi(t) = \tan^{-1}\left(\frac{1}{\cos\Delta\phi}\frac{V}{(V+\Delta V)}\frac{v_I(t) - V_{OSI}}{v_Q(t) - V_{OSQ}} - \tan\Delta\phi\right) \tag{4.30}$$

The measured I and Q signals are typically obtained through an average of many measured values to cancel out the noise components, which are composed of the phase noise of the LO source and the white noise generated by components in the system.

4.2.2 Millimeter-Wave SFCW Radar Sensor Transceiver

Figure 4.3 shows the block diagram of the 29.72–37.7-GHz millimeter-wave SFCW radar sensor based on the coherent super-heterodyne architecture. It is a mono-static system, transmitting and receiving signals through the same antenna. The sensor is mono-static, transmitting and receiving signals through the same antenna. It consists of a transceiver, a waveguide horn antenna, and digital signal processing embedded in LabView.

The 29.72–37.7-GHz millimeter-wave SFCW radar sensor is operated as follows. A sinusoidal signal of 1.72 GHz is generated by a phase-locked loop (PLL) oscillator consisting of a temperature-compensated crystal oscillator (TCXO) having a reference frequency of 6.71875 MHz, a 2^8-frequency divider and a 2-pole loop filter. The generated CW is used as an IF signal at the sub-harmonically pumped mixer in the transmitter path and as a LO signal at the quadrature detector in the receiver path. The sub-harmonically pumped mixer up-converts the IF signal, modulated with the incoming 14–17.99-GHz stepped frequencies with 10-MHz frequency increments from an external synthesizer, to 29.72–37.7-GHz signals that are transmitted toward a target. The reflected signals from the target are down-converted to a single IF signal of 1.72 GHz by mixing with the 14–17.99-GHz stepped-frequency signals in the sub-harmonically pumped mixer in the receiver path, which is converted into the base-band I and Q signals by the quadrature detector. These I and Q signals are then digitized with ADCs in the DAQ board of LabView and processed to extract the target's information.

The transceiver is completely realized using microwave and millimeter-wave integrated circuits - both MICs and MMICs. It is separated into two parts, one for high-frequency circuits and another one for low-frequency circuits, for easy fabrication, evaluation and trouble-shooting, as well as facilitating circuit design.as shown in Fig. 4.3. The high frequency circuits are integrated on an alumina substrate having a thickness of 0.0254 cm and a relative dielectric constant of 9.8. The low-frequency circuits are realized on a low-cost FR-4 substrate that has a thickness of 0.0787 cm and a relative dielectric constant of 4.3. Figure 4.4 shows a

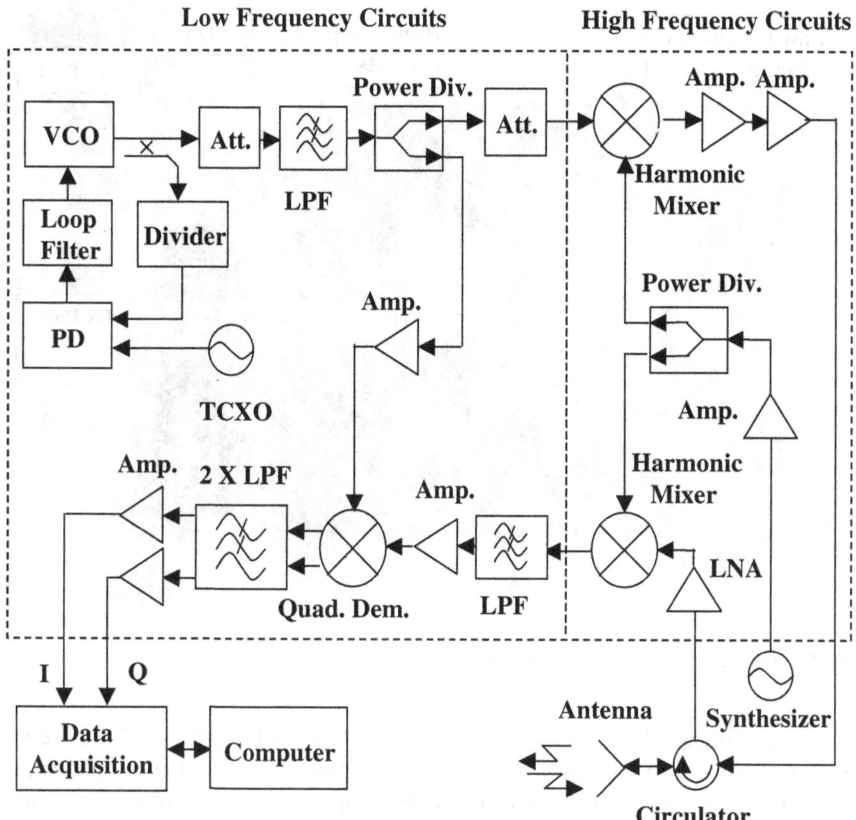

Fig. 4.3 Block diagram of the 29.72–37.7-GHz millimeter-wave SFCW radar sensor

photograph of the transceiver. The alumina and FR-4 substrates are mounted on an aluminum block, which acts a common ground plane for the high- and low-frequency circuits and supports the transceiver. The transceiver has overall dimensions of 10.2 × 15.2 cm.

The high frequency circuits include a Ku-band medium-power amplifiers (Agilent, HMMC-5618), Ka-band sub-harmonically pumped mixers (Hittite, HMC266), Ka-band low-noise amplifiers (TRW ALH140C) and Ku-band power divider, integrated together through microstrip lines on an alumina substrate. All the components are bonded with 0.00762 by 0.00127 cm gold ribbons using a wedge-bonding machine. The Ku-band medium-power amplifier increases the external LO power generated from the synthesizer. The harmonic mixer up-converts the input signal at the IF port (for transmitting) or down-converts the input signal at the RF port (for receiving) with the second harmonic of the signal at the LO port. The LNAs amplify the receiving signals and maintain a low noise figure for the receiver.

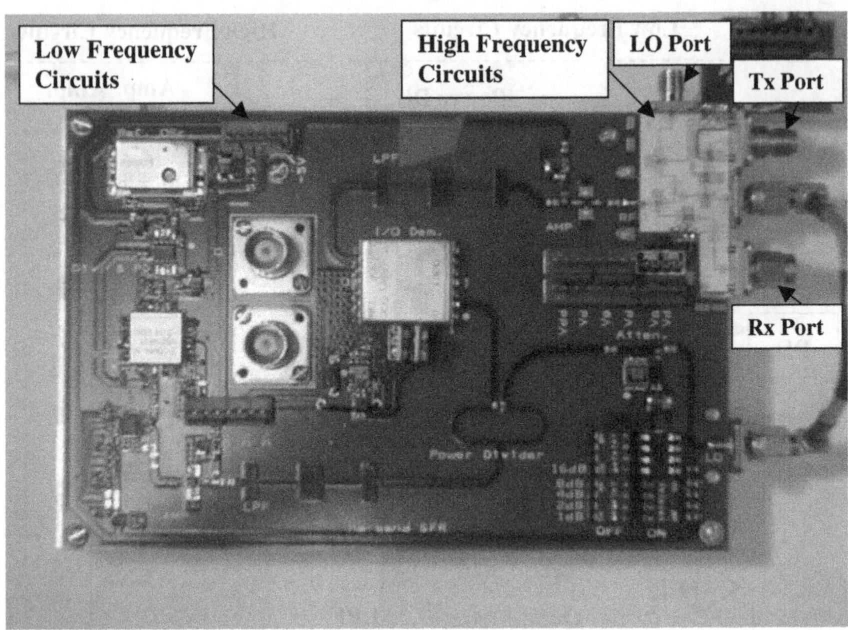

Fig. 4.4 Photograph of the millimeter-wave SFCW radar sensor's transceiver

The low-frequency circuits operate below 1.72 GHz and consist of a PLL circuit, two attenuators, two low-pass filters (LPFs), a power divider, two amplifiers, a quadrature detector and a two-channel video amplifiers, which are mounted and etched on a FR-4 substrate. The PLL oscillator generates a stable single IF and the attenuators adjust the LO and IF power below the specifications of the following circuits. The LPFs reduce the high-frequency harmonics contained in the IF signal and the IF harmonics added in the base-band I/Q signals. The power divider splits the IF signal into two IF signals: one for the IF signal of the up-converter and the other for the LO signal of the quadrature detector. The LO amplifier increases the LO power sufficiently to pump the quadrature detector and the quadrature detector down-converts the single-frequency input signal, which includes the target's information, into the base-band I/Q signals. The two-channel video amplifier increases the power of the base-band I/Q signals to meet the required input power range of the ADC.

A waveguide horn antenna with gain within 23.7–24.7 dBi and half-power beam-widths of about 15° and 8° at the E- and H-plane, respectively, over the sensor's operating frequencies was employed. The same antenna was used for both transmitting and receiving signals via a circulator.

The base-band I and Q signals are sampled through a 12-bit ADC in a data acquisition module of LabView. The resultant digitized I and Q signals are then formed into an array of complex signals, which are then transformed into a time-compressed waveform called synthetic pulse using IDFT algorithm. The synthetic pulse contains the target information. Zero-padding was used to reduce the error in range and improve the speed of IDFT. In this signal processing, we used a Hamming window, 4096 points of IDFT and a frequency step of 20 MHz, which result in ±0.09 cm error in range. This signal processing was done based on the signal processing described in Sects. 4.2.5 and 4.2.6.

Table 4.1 shows the results of the analysis of the receiver. The transmission loss L_t = 13 dB was measured with a waveguide horn antenna, as shown in Fig. 3.15, and used to estimate the available transmitting and receiving power. The LNA and down-converter need to be operated below their respective 1-dB power compression points to avoid saturation. To that end, it is required that the maximum available receiving power input to the LNA must be below −7 dBm. This power level results in 4 dBm at the output of the LNA that has 11-dB gain, which is below the 4-dBm input power compression of the down-converter. The maximum available receiving power is then set to −8 dBm to provide a 1-dB margin.

A total noise figure of 5.7 dB is estimated from the noise figures and gains of the LNA (4 dB, 11 dB) and down-converter (12 dB, −12 dB). When the synthesizer's fastest *PRI* of 100 ms, resulting in an instantaneous bandwidth of 10 Hz, is used with the output *S/N* set to 14 dB, the receiver sensitivity S_R is calculated as −150.3 dBm from (2.63).

The input voltage range of the ADC is determined according to the ADC's specifications. The ADCs in the DAQ board have 12 bits of resolution per sample, which results in a dynamic range of 72 dB, and the maximum input range is between ±0.2 V to ±42 V, which leads to a sensitivity of 35 μV at ±0.2 V of the maximum input signal. The maximum input signal is then chosen to be ±2 V, which results in the input voltage range of the ADC from ±2 V (or 9 dBm at 1 KΩ) to ±0.5 mV (or −63 dBm). The video amplifier is used for boosting the quadrature detector's output to the ADC's input voltage range.

Table 4.1 Receiver's analysis where $P_{in,1dB}$ is the input 1-dB compression point and P_{out} is the output power

	Gain (dB)	Loss (dB)	$P_{in,1dB}$ (dBm)	P_{out} (dBm)
LNA	11		4	3
Down-converter		12	4	−9
LPF		0.5		−9.5
Amplifier	13		1	3.5
I/Q mixer		8	4	−4.5
LPF (Ro = 200 Ω)		6.2		−10.7
Amplifier (Ro = 1 KΩ)	27.7			10
FR-4 Substrate		1		9
Total	51.7	27.7		
Receiver's available dynamic range $DR_{R,max}$		72 dB	Actual system performance factor SF_a	108 dB

The maximum available receiving power $P_{r,max}$ is set to −8 dBm. 1-dB insertion loss for the FR-4 substrate is assumed

Table 4.2 Transmitter's analysis

	Gain (dB)	Loss (dB)	$P_{in,1dB}$ (dBm)	P_{out} (dBm)
PLL Oscillator				5
Attenuator		2		3
LPF		0.5		2.5
Splitter		3.5		1
Attenuator		5		−4
Up-converter		12	4	−16
Amplifier	11		4	−5
Amplifier	11		4	6
FR-4 Substrate		1		7
Total	22	24		

The insertion loss of the FR-4 substrate is assumed to be 1 dB

From the maximum available receiving power of −8 dBm, the maximum available transmitting power is calculated as 5 dBm from (3.37). The system performance factor *SF* is then determined as 155.3 dB (=5 + 150.3 dBm) using (3.38). With the maximum available receiving power of −8 dBm, the receiver's maximum available dynamic range can be calculated from (3.39) as 142.3 dB (=−8 + 150.3 dBm). However, the ADC's dynamic range limits the receiver's available dynamic range $DR_{R,max}$ to 72 dB. Therefore, the actual system performance factor, SF_a, for the radar equation is determined from (3.43) as 108 dB (=95 + 13 dB).

Similarly, Table 4.2 shows the results of the analysis of the transmitter. In order to reach the maximum available transmitting power level of 5 dBm, two cascaded amplifiers are used. Two attenuators are used to tune the power level, one between the PLL oscillator and the splitter's output and the other between the splitter's output and the up-converter's input.

4.2.3 *Microwave SFCW Radar Sensor Transceiver*

Figure 4.5 shows the system block diagram of the 0.6–5.6 GHz microwave SFCW radar sensor based on the coherent super-heterodyne architecture. This frequency range is chosen to allow sufficient penetration while achieving good resolution. The sensor consists of a transceiver, two antennas and a digital signal processing embedded in LabView.

The temperature compensated crystal oscillator (TCXO) in the transceiver generates a signal of 10 MHz, which is used as the LO signal for the quadrature detector and the IF signal for the up-converter. The up-converter converts the incoming 0.59–5.59-GHz LO signals from the synthesizer to 0.6–5.6-GHz signals to be transmitted toward targets (through the ultra-wideband transmit antenna.) Alternately, the down-converter converts the returned signals from targets (through the receive antenna) to an IF signal of 10 MHz by mixing them with the coherent

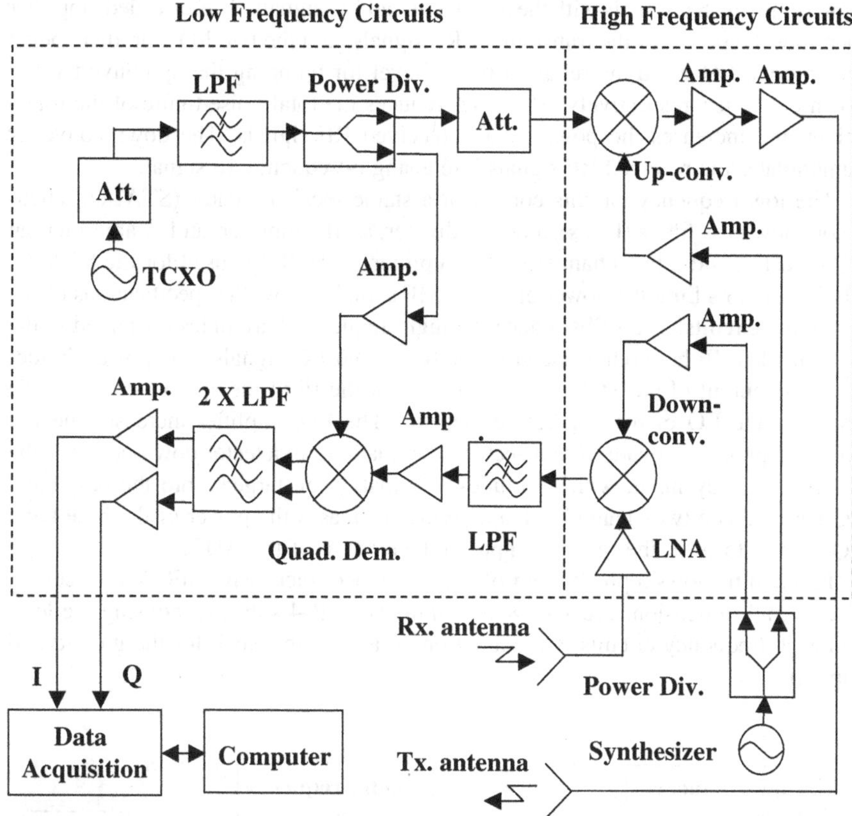

Fig. 4.5 Block diagram of the microwave SFCW radar sensor

LO signals from the synthesizer. The IF signal is then converted into the base-band I/Q signals in the quadrature detector by mixing it with the coherent LO signal from the TCXO. The I/Q signals are finally digitized with ADCs and processed in digital signal processing blocks to extract the target information.

The transceiver is realized using microwave integrated circuits (MICs). As for the millimeter-wave SFCW transceiver described in Sect. 4.2.2, the microwave SFCW transceiver is separated into two parts for easy fabrication, evaluation, and trouble-shooting. One is for low-frequency circuits and the other is for high-frequency circuits. Both low and high-frequency circuits are fabricated on the same 0.7874–mm FR-4 substrate for low cost and convenience integration. To alleviate the relatively high loss of the FR-4 substrate (about 0.1575 dB/cm at 5 GHz) as compared to the commonly used RT/Duroid substrates for microwave circuits, the size of the high-frequency circuits is designed in a compact size of 5 × 10 cm.

The high-frequency circuits include an up-converter, a cascaded RF amplifier, two LO amplifiers, a LNA, and a down-converter. The up-converter modulates the IF signals into the RF signals with the aid of external LO signals. The cascaded amplifier increases the power of the transmitting RF signals, and the two LO amplifiers boost the external LO up to the required power level for pumping the up-converter and down-converter, respectively. The LNA reduces the total noise figure of the transceiver and increases the power of the received RF signals. The down-converter demodulates the received RF signals into a single-frequency IF signal.

The low-frequency circuits consist of a stable local oscillator (STALO), attenuators, low pass filters (LPFs), a power divider, an IF amplifier, an LO amplifier, an I/Q detector, and a two-channel video amplifier. A TCXO is used for the STALO. The attenuators limit the power of LO and IF signals below the specifications of the following circuits. The LPFs reduce the high frequency harmonics contained in the IF signal and IF harmonics added in the base-band I/Q signals. The power divider splits the output of the TCXO into two, one for the IF of the up-converter and the other for the LO of the quadrature detector. The LO amplifier increases the LO power to pump the quadrature detector. The quadrature detector down-converts the single-frequency input, which includes information on targets, into the base-band I/Q signals. The two-channel video amplifier increases the power of the base-band I/Q signals to meet the required input voltage range of the ADC.

Figure 4.6 shows a photograph of the developed microwave SFCW transceiver. The overall dimensions are 10.2 × 17.8 mm. The FR-4 substrate housing the low- and high-frequency circuits is mounted on an aluminum block for the ground and support.

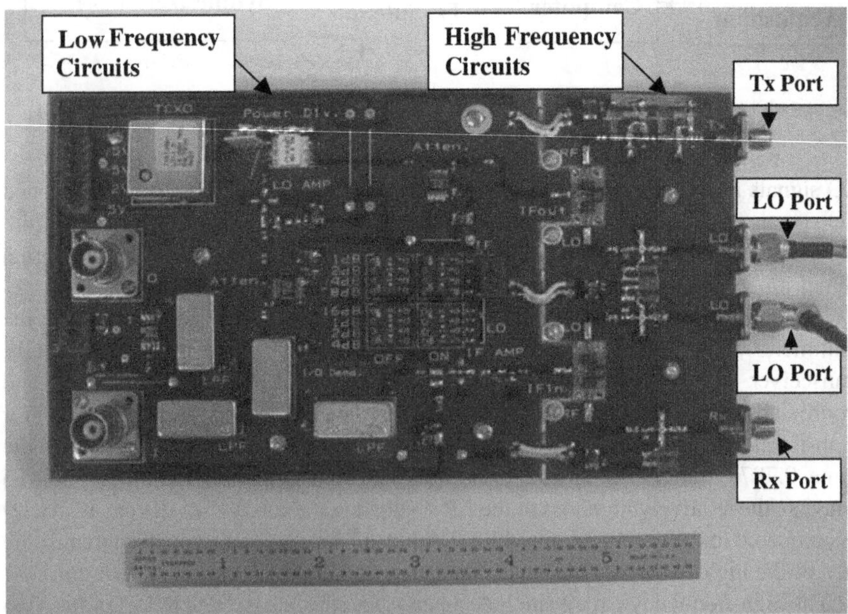

Fig. 4.6 Photograph of the microwave SFCW transceiver

Table 4.3 Results of transmitter's analysis at 3 GHz. A 1-dB insertion loss is assumed for the employed FR-4 substrate

	Gain (dB)	Loss (dB)	$P_{in,1dB}$ (dBm)	P_{out} (dBm)
STALO				5
Attenuator		3		2
LPF		0.3		1.7
Splitter		3.2		−1.5
Attenuator		2.5		−4
Up-converter		8	5	−12
1st Amplifier	12		3	0
2nd Amplifier	12		3	12
FR-4 Substrate		1		11

Table 4.3 shows the results of the analysis of the transmitter design at 3 GHz. The transmission loss L_t of 25 dB was first measured with the developed antennae to estimate the available transmitting and receiving power. The maximum available transmitting power is set to 11 dBm to avoid the transmitter's amplifier from saturating. Two attenuators are used for adjusting the power levels, one between the outputs of the STALO and splitter and another between the splitter's output and the up-converter's input.

Table 4.4 shows the results of the analysis of the receiver design at 3 GHz. From the maximum available transmitting power, the maximum available receiving power is estimated as −14 dBm from (3.37). The input voltage range of the ADC is set to ±2 V (or 9 dBm at 1 KΩ), which is the same as that used for the millimeter-wave SFCW radar sensor system, and the video amplifier is used to increase the quadrature detector's output level to the ADC's required input range.

Table 4.4 Results of the receiver's analysis at 3 GHz

	Gain (dB)	Loss (dB)	$P_{in,1dB}$ (dB)	P_{out} (dB)
LNA	12		3	−2
Down-converter		8	5	−10
LPF		0.3		−10.3
Amplifier	13			2.7
I/Q mixer		6	4	−3.3
LPF (Ro = 200 Ω)		6.2	–	−9.8
Amplifier (Ro = 1 KΩ)	26.8		–	10
FR-4 substrate		1		9
Receiver's available dynamic range $DR_{R,max}$		72 dB	Actual system performance factor SF_a	121 dB

The maximum available receiving power is assumed to be −8 dBm and the insertion loss of the FR-4 substrate is assumed as 1 dB

A total noise figure of 6 dB is estimated from the noise figures and gains of the LNA (5.5, 12 dB) and down-converter (8, −8 dB). When the other conditions are the same as those for the millimeter-wave SFCW radar sensor, the receiver sensitivity S_R is estimated as –148 dBm from (2.63).

From the maximum transmitting power of 11 dBm, the system performance factor *SF* is calculated as 159 dB (=11 + 148 dBm) by using (3.38). The receiver's maximum available dynamic range is determined as 134 dB (=−14 + 148 dB) from (3.39). With the ADC's dynamic range of 72 dB, the actual system performance factor for the radar equation is calculated as 121 dB (=96 + 25 dB) from (3.43).

4.2.4 Antennas

Wideband radar sensors employ various broadband antennas such as log-periodic, spiral, waveguide horn, and transverse electromagnetic (TEM) horn antennae. The log-periodic antenna has good polarization and broad bandwidth; however, its physical size restricts its use drastically. The spiral antenna has a wide bandwidth, but its use is also limited due to its dispersive characteristics. The waveguide horn antenna can operate only within the waveguide's operating frequency range and is highly dispersive and expensive to manufacture. The TEM horn antennas are attractive for ultra-wideband radar sensors owing to their inherent characteristics of wide bandwidth, high directivity, good phase linearity, and low distortion. Various types of TEM horn antennas have been developed [39–41]. A TEM horn antenna, however, needs a balun at its input, prohibiting a direct connection between antennas and the transceiver circuit. The use of balun also limits the antenna's operating bandwidth. Moreover, direct coupling between the transmitting and receiving antennas in a bi-static system that uses two antennas closely spaced is severe. The TEM horn antennas are also relatively large and costly to build.

Microstrip quasi-horn antennas that possess an extremely broad bandwidth of multiple decades, relatively high gain, and compatibility with microstrip circuits, were developed and demonstrated up to Ka-band (26.5–40 GHz) GHz [42–44]. The microstrip quasi-horn antennas are suitable for the microwave and millimeter-wave SFCW radar sensors. The microstrip quasi-horn antennas have similar performance compared to the waveguide horn antennas, but they can operate over wider bandwidths, do not need a transition to printed circuits, and are much easier to produce at a much lower cost. As compared to the TEM horn antennas, the microstrip quasi-horn antennas have a smaller size and allow a direct integration with the microstrip-based transceivers while maintaining adequate isolation between the two antennas when placed next to each other.

Figure 4.7 shows a sketch of the microstrip quasi-horn antenna. It consists of a conductor on top of a grounded dielectric substrate and hence resembling a microstrip structure. The dielectric medium, and hence the height of the conductor above the ground plane, can be changed in any particular fashion, and the conductor's profile depends on the contour of the dielectric substrate. The microstrip

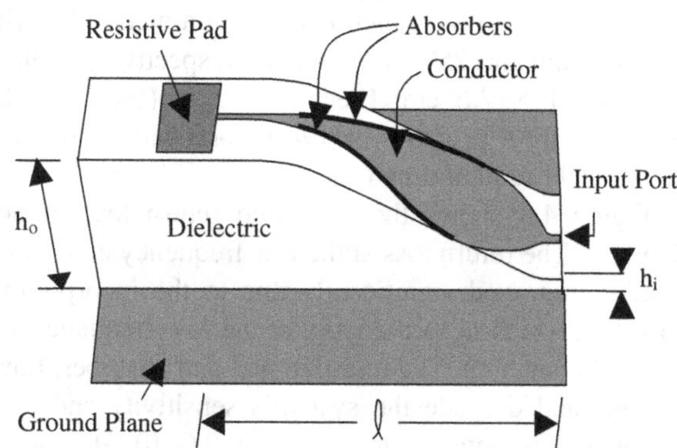

Fig. 4.7 Sketch of the microstrip quasi-horn antenna. The resistive pad and absorber are optional and used as appropriate

quasi-horn antenna can be directly connected to the connector without any balun and/or transition, making it simpler physically and higher-performance electrically. When two such antennas used for the transmitter and receiver are placed against each other, the common ground plane acts as a shield between these antennas, resulting in a high isolation between them. This unique feature is extremely attractive for radar sensor applications where a certain degree of isolation is always needed between transmit and receive antennas.

The operation of the microstrip quasi-horn antenna is based on the principle of wave propagation along a transmission line. In the uniform section of the microstrip line, where the spacing between the top conductor and the ground plane is very small compared to a wavelength, wave propagation is mostly confined within the dielectric between the top conductor and the ground plane. However, as the separation between the conductor and ground plane gradually increases and approaches approximately a half-wavelength or more, the energy begins to radiate in the end-fire mode and, consequently, the wave is no longer guided between the conductor and ground plane. The entire structure effectively behaves as an antenna. The width of the microstrip quasi-horn antenna aperture primarily controls the radiation at low frequencies and hence sets the low-frequency limit for radiation, while the antenna length and the conductor and dielectric contours control the matching over the operating bandwidth.

4.2.4.1 0.5–10 GHz Microstrip Quasi-Horn Antenna

The microstrip quasi-horn antenna for the microwave SFCW radar sensor is designed to present at least 10 dB of return loss over a wide band of 0.5–10 GHz. The length of the antenna, which is primarily restricted by the lowest operating frequency, was set to 40.6 cm. Styrofoam, which has nearly the same relative dielectric constant ($\varepsilon_r = 1.03$) as air, is used as the dielectric medium to support the antenna's top conductor. Reflections from the open end and the edges are significantly reduced by appending a resistive pad to the open end and absorber to the edges as illustrated in Fig. 4.7.

The resistive pad, which is made of a metal film with a thickness and resistivity of 0.635 mm and 250 Ohms/square, respectively, is tuned empirically to an optimal size of 5.1 × 7.6 cm. Electromagnetic (EM) simulations are performed using Ansoft's HFSS program [45] to theoretically verify the reflection coefficients and the far field radiation patterns.

Figure 4.8 shows the measured return loss in both the time and frequency domains. The return loss at the low frequency end, as seen in the frequency-domain plot, is improved significantly due to the incorporation of the resistive pad and absorber, which absorb energy at the low frequencies that cannot be radiated by a finite-size antenna. The resistive pad and absorber, however, reduce the gain of the antenna and degrade the system's sensitivity and dynamic range. The measured return loss is better than 12 dB at 0.6–10-GHz as shown in Fig. 4.8. A better illustration of the impact of these accessories is shown in the time-domain plots. An

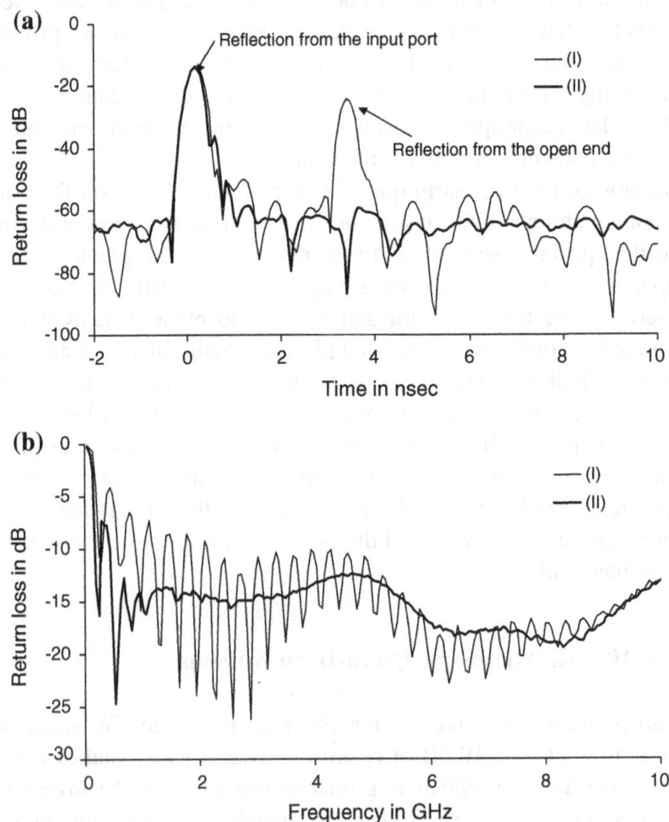

Fig. 4.8 Antenna's return loss in time domain (**a**) and frequency domain (**b**), where (I) indicates the antenna alone and (II) represents the antenna with resistive pad and absorber

additional narrow peak, indicating deterioration of the input reflection loss, is observed at around 3.5 ns when the resistive pad and absorbers are not incorporated.

Figure 4.9 shows the simulated radiation patterns in the E-plane. The simulated gain and the 3-dB beamwidth are within 6–17 dBi and 25°–45° at 0.6, 3, and 5 GHz, respectively.

The simulated results show that E-plane patterns are tilted about 8°–28° off the boresight axis due to the ground-plane effect. Therefore, in practical use, the transmitting and receiving antennas should be carefully aligned to achieve maximum possible gains. As an example illustrating a possible way to optimize the alignment of two antennas, measurements were performed on a pavement sample consisting of asphalt, base and subgrade using a network analyzer. Figure 4.10 shows the configuration of the aligned antennas above the pavement sample. A set of initial values for the angle, stand-off distance, and gap are obtained from EM

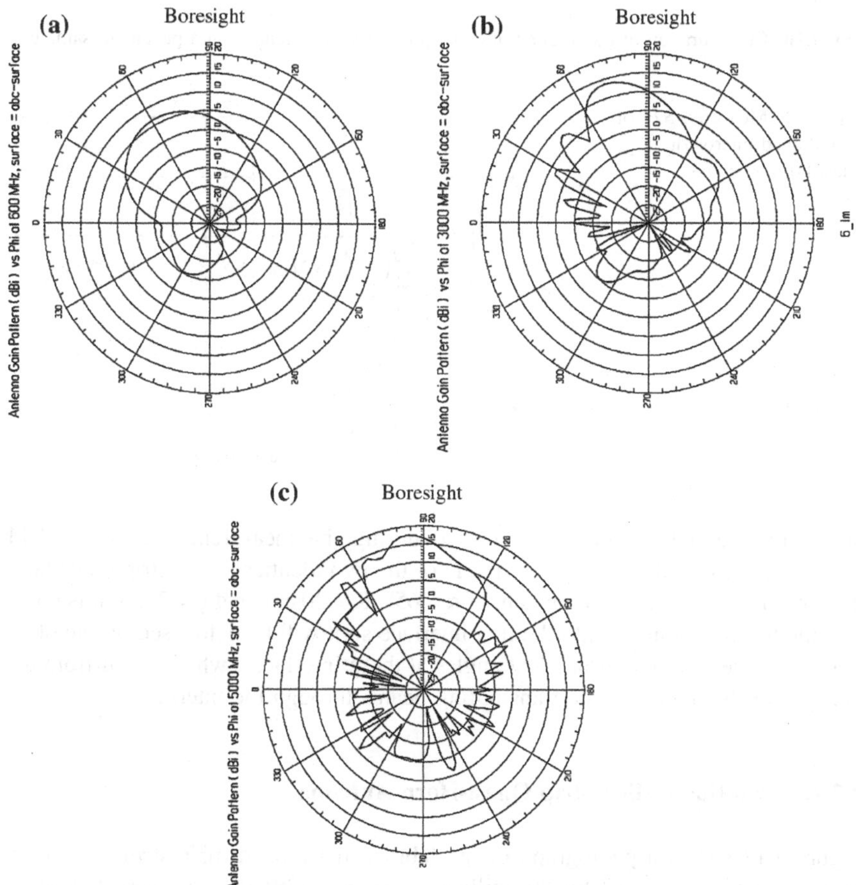

Fig. 4.9 Calculated E-plane radiation patterns at 0.6 GHz (**a**), 3 GHz (**b**), and 5 GHz (**c**)

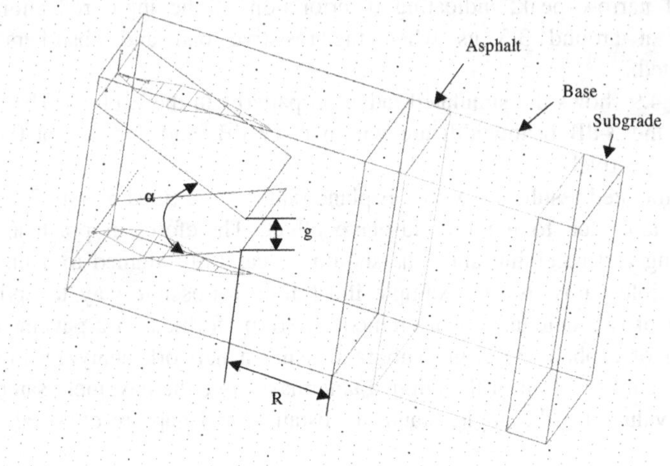

Fig. 4.10 Configuration of the aligned microstrip quasi-horn antennas on a pavement sample

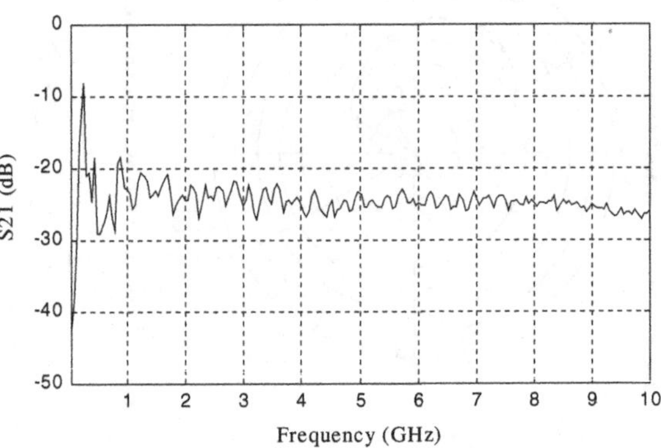

Fig. 4.11 Measured S21 of two aligned microstrip quasi-horn antennas

simulations, and then they are adjusted during the measurements. Figure 4.11 shows the measured insertion loss (S21) of the two identical microstrip quasi-horn antennas for the optimum alignment of $\alpha = 65°$, R = 20 cm, and g = 7 cm. It is noted that the beam is pointed onto the first interface so it will be defocused on the other interfaces due to different incident angles at these interfaces, which result from the changes of the angles of the transmitting beams through the interfaces.

4.2.4.2 Ka-Band Microstrip Quasi-Horn Antenna

Figure 4.12 shows a photograph of the fabricated Ka-band microstrip quasi-horn antenna that can be used for the millimeter-wave SFCW radar sensor. As can be seen, the antenna is directly connected to the connector without a balun, which

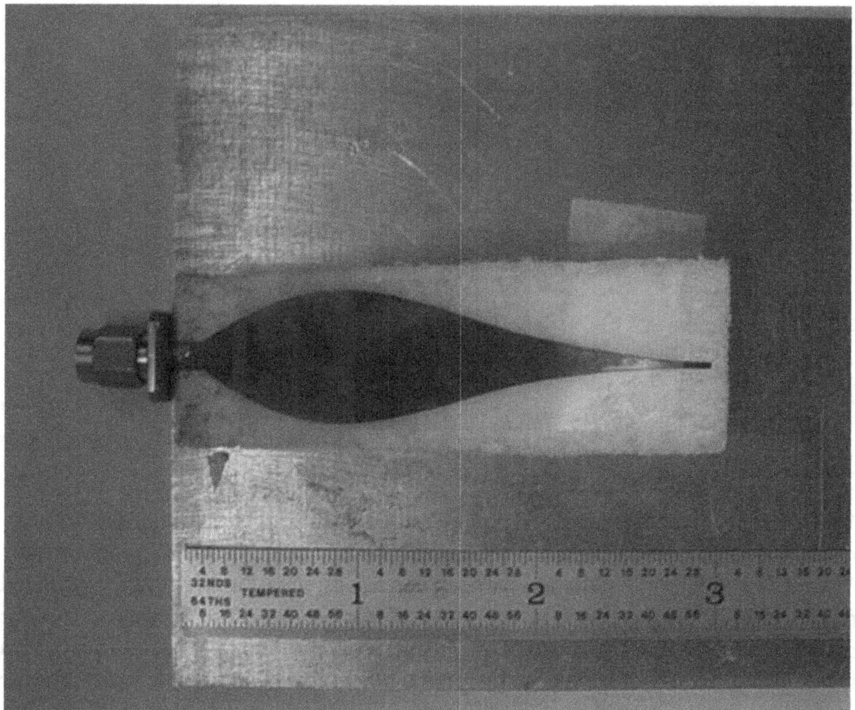

Fig. 4.12 Photograph of the Ka-band microstrip quasi-horn antenna. No resistive pad and absorber are used

results in simple structure physically and good performance electrically. To improve both the impedance transformation and the antenna radiation characteristic, especially at lower frequencies, the antenna is shaped according a combination of the exponential and cosine-squared functions to determine the height between the top conductor and the ground plane. This combination is selected by comparing the simulation results for different shapes of the antenna. The exponential function is used to determine the height up to one-half wavelength of the lowest frequency (26.5 GHz) and the cosine-squared function is used for the height from one-half wavelength to the open end. The impedance change from the input port to the open end follows that of an exponential taper.

Figure 4.13 shows the measured return loss of the Ka-band microstrip quasi-horn antenna, which is better than 14 dB from 20 to 40 GHz. Figure 4.14 shows the calculated and measured H-plane radiation patterns at 26.5 and 35 GHz, where the patterns are measured and calculated within −90° to +90° from the boresight. The calculated and measured gains are within 16–18 dBi and 14.5–15 dBi, respectively. Both the computed and measured half-power beamwidths are less than 20°. Figure 4.15 shows the calculated and measured E-plane radiation patterns at 26.5 and 35 GHz. The calculated and measured gains are within 16–18.5 dBi and

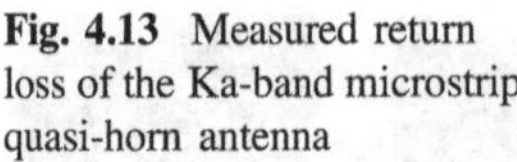

Fig. 4.13 Measured return loss of the Ka-band microstrip quasi-horn antenna

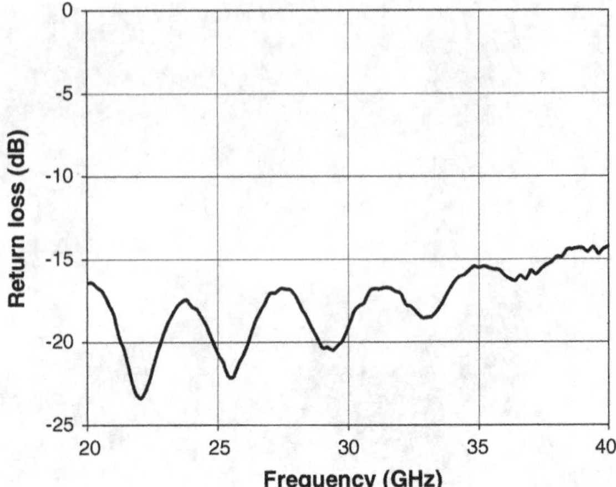

14.5–15.5 dBi, and the calculated and measured beamwidths are about 22° and 15°, respectively. All the calculations are carried out using Ansoft HFSS program [45]. The measured H-plane radiation patterns agree reasonably well with the calculated results, despite some errors in the physical dimensions. The measured E-plane radiation patterns show a reasonable agreement with those calculated but having a lower gain primarily due to the finite ground plane. It was impossible to achieve the same dimensions and shapes for the fabricated antenna as those used in the simulations because the top conductor and foam were cut and integrated manually. It is important to note that both the measured and simulated radiation patterns of the E-plane are tilted about 10° off the bore-sight axis due to the ground-plane effect.

4.2.5 Signal Processing

The base-band analog I/Q signals are digitized into digital I/Q signals at ADCs in the DAQ board used with Labview. These digitized I/Q signals need to be processed by a signal processing to transform into synthetic pulses in the time domain. To that end, the signal processing including I/Q error compensation and IDFT was developed for the microwave and millimeter-wave SFCW radar sensors using LabView.

The returned stepped-frequency from a target must be synchronized during data acquisition. That is, the starting point of the data acquisition should coincide with the desired first step frequency. To synchronize the first step frequency to the starting point of the data acquisition, a trigger input is used, and Labview needs to be programmed to utilize the trigger input for synchronizing data acquisition.

The I/Q data collected through the data acquisition is processed to compensate for the I/Q errors. This compensation is focused on the I/Q errors caused by the common amplitude and phase errors. The differential amplitude and phase errors

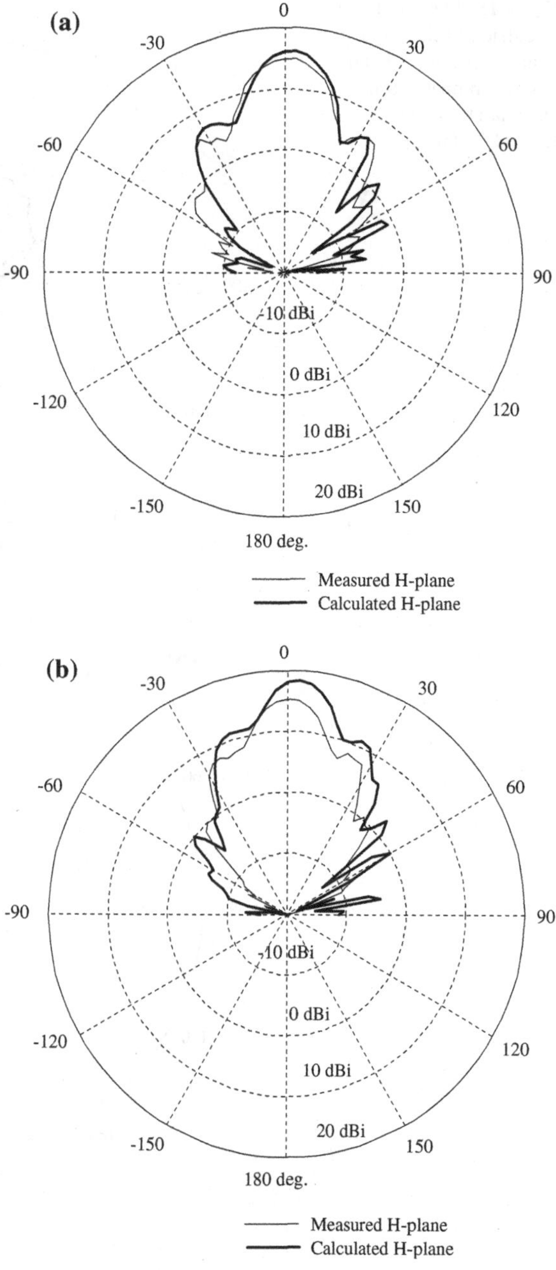

Fig. 4.14 Measured and calculated H-plane radiation patterns of the Ka-band microstrip quasi-horn antenna at 26.5 GHz (**a**) and 35 GHz (**b**)

can be easily compensated in the super-heterodyne scheme [46]. These compensated I/Q signals are then incorporated into the complex vectors $C_i = I_i + jQ_i$, $i = 0, 1, 2, \ldots, N - 1$, described in (3.5) and $(M - N)$ zeros are added onto these complex

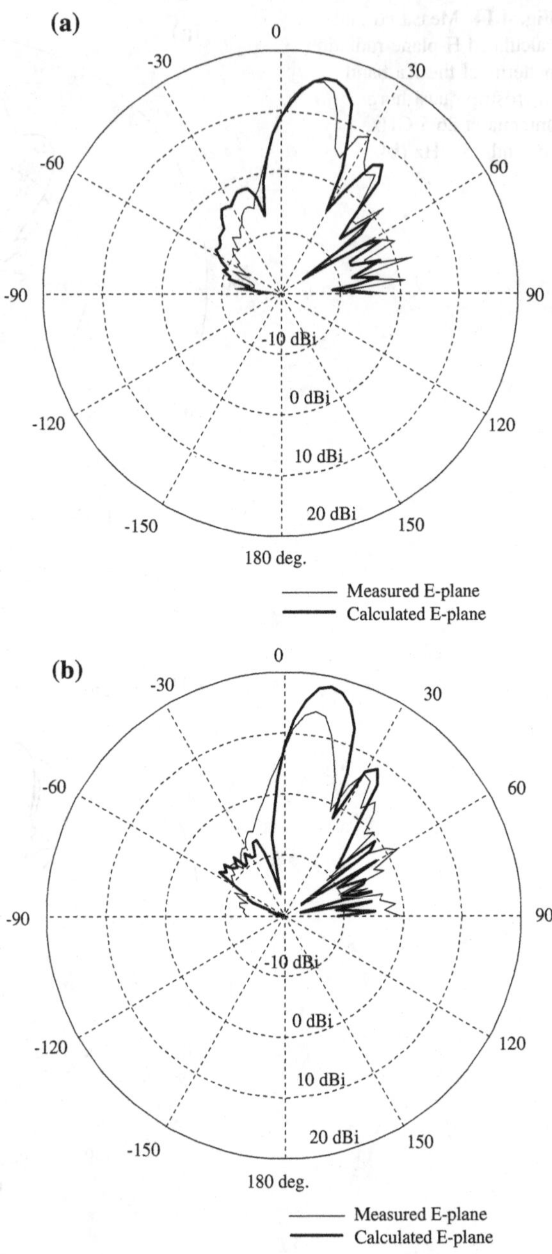

Fig. 4.15 Measured and calculated E-plane radiation patterns of the Ka-band microstrip quasi-horn antenna at 26.5 GHz (**a**) and 35 GHz (**b**)

vectors before combining them into the complex vector array $V = [C_0, C_1, \ldots, C_{N-1}]$ described in (3.6). Finally, the vector array is imposed with a Hamming window to reduce the side-lobes and transformed to a synthetic pulse in the time domain.

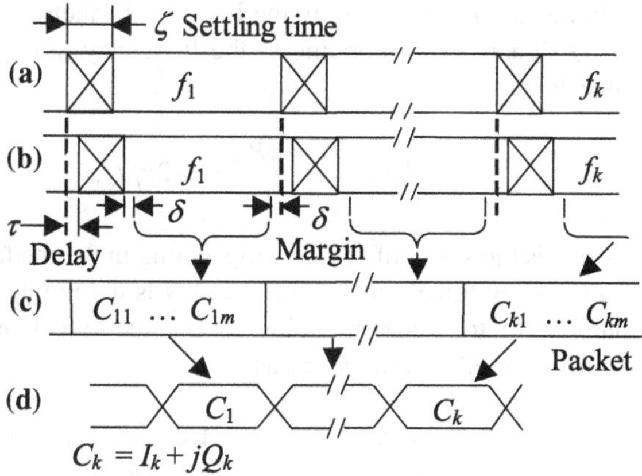

Fig. 4.16 Procedure for generating representative complex vectors: **a** transmitted signals, **b** received signals, **c** restored effective complex vectors, and **d** representative complex vectors after averaging

The digitized I and Q samples from ADCs needed to be synchronized, restored, filtered and averaged to obtain a representative data point. Figure 4.16a, b represent the train of the frequency steps of the transmitted and returned signals, respectively. The settling time, ζ, of the synthesizer, and the delayed time, τ, of the received signals should be considered since non-coherent demodulation occurs during the time $\zeta + \tau$. The samples are useless during the time $\zeta + \tau$, and it is thus necessary to restore the samples with effective samples only.

The effective samples shown in Fig. 4.16c are reconstructed such that there are adequate safety margins (δ) to ensure that each packet has valid samples. The samples in each packet, $C_{k1}, C_{k2}, \ldots, C_{km}$ ($k = 0, 1, 2, \ldots, N - 1$) where C_{km} denotes the mth complex vector corresponding to the kth frequency, are filtered and averaged to generate a new complex vector, C_k, as shown in Fig. 4.16d. Averaging the samples reduces the errors caused by short time jitters of the TCXO and synthesizer.

4.2.6 Compensation for the I/Q Errors

A practical system produces common and differential amplitude and phase errors in the I and Q channels. The common error is the error caused by common circuits in the signal-propagation path to both the I and Q channels, which consist of antennae, amplifiers, other mixers, transmission lines, filters, etc. The differential error is caused by the mismatch between the I and Q channels; it is the main error and is the commonly known I/Q error. The differential error is generated in the quadrature detector due to the difference between the I and Q mixers and the phase imbalance of the 90° phase-shift circuit (e.g., 90° coupler) contained in the quadrature detector. For a super-heterodyne scheme, the differential amplitude and phase errors in the I and Q channels are normally constant over the RF band of interest due to a (down-converted) single constant intermediate frequency (IF) as discussed earlier.

In the absence of errors in the I and Q channels, the phase ϕ_i of the base-band I and Q signals, corresponding to the frequency f_i, expressed in terms of the target range R is

$$\phi_i(R,f_i) = -\frac{4\pi R f_i}{v} = -2\pi f_i t_d, \quad i = 0, 1, \ldots, N-1 \tag{4.31}$$

where v is the speed of signals propagating in the medium, t_d is the time delay equal to a two-way travel time of $2R/v$, and N is the number of the frequency steps. The complex vectors corresponding to a fixed target are expressed in terms of f_i, assuming equal amplitude A_i, as

$$\begin{aligned} I_i(f_i) + jQ_i(f_i) &= A_i \cos[\phi_i(f_i)] + jA_i \sin[\phi_i(f_i)] \\ &= A_i e^{-j2\pi f_i t_d} \end{aligned} \tag{4.32}$$

If the common and differential errors are included, the complex vectors become, assuming $A_i = 1$ for simplicity without loss of generality:

$$\begin{aligned} I_i(f_i) + jQ_i(f_i) = {} & (1 + \frac{cg_i}{2}) \cos[2\pi f_i t_d + cp_i] \\ & - j(1 + dg_i + \frac{cg_i}{2}) \sin[2\pi f_i t_d + dp_i + cp_i] \end{aligned} \tag{4.33}$$

where cg_i, cp_i and dg_i and dp_i are the common and differential amplitude and phase errors, respectively.

The differential amplitude and phase errors generate a Hermitian image of the response in the resultant synthetic range profile, resulting in a reduction of the sensor's unambiguous range by one-half [46]. In a super-heterodyne system, these errors are constant across the RF operating frequency range since a single constant IF is typically used for the quadrature detector. Consequently, the measurement and compensation of these errors is simple. The differential amplitude and phase errors in the I and Q channels at IF can be measured by using the methods presented in [23, 46]. By following these techniques, the differential amplitude and phase errors were measured as 1 dB and 3° and 3.5 dB and 7° for the microwave and millimeter-wave SFCW radar sensors, respectively.

The common phase error can be described as consisting of a linear phase error $2\pi f_i \alpha$ and a non-linear phase error β_i as

$$cp_i = 2\pi f_i \alpha + \beta_i \tag{4.34}$$

The common linear phase error results in a constant shift of the output response's synthetic range profile due to the fact that a frequency-dependent linear phase is transformed into a constant time delay through the Inverse Fourier Transform [37]. Therefore, it is not necessary to correct the common linear phase error. On the other hand, the non-linear phase error causes shifting as well as imbalance in the response of the synthetic range profile. The common amplitude error also affects the response, e.g., the shape of the synthetic range profile, as they tend to defocus the response in the profile and increase the magnitudes of side lobes. Therefore, these common non-linear phase and amplitude errors need to be corrected. The following formulates a simple, yet effective and accurate, technique for compensation of these errors.

The complex vector given in Eq. (4.33) for a fixed frequency f_k is rewritten in terms of the range R as

$$\begin{aligned} I(R) + jQ(R) &= (1 + \frac{cg_k}{2})\cos[2\pi f_k t_d(R) + cp_k] \\ &\quad - j(1 + dg_k + \frac{cg_k}{2})\sin[2\pi f_k t_d(R) + dp_k + cp_k] \end{aligned} \tag{4.35}$$

where $t_d(R)$ signifies the dependence of t_d on R. It is then seen that these complex vectors will rotate circularly if the I and Q channels are completely balanced when R is increased or decreased at a constant rate. In the process of correction, the complex vector, $I(R) + jQ(R)$, is measured when a metal plate is moved along a track at a fixed frequency. The metal plate has a size of 3-by-3-foot in order to accommodate the sensor's lateral resolution. Initially, the complex vector rotates elliptically, either clockwise or counter-clockwise, with respect to the direction of the metal plate, as the I and Q components are not orthogonal due to the differential phase errors. After these differential errors are corrected, Eq. (4.35) becomes

$$\begin{aligned} I(R) + jQ(R) &= (1 + \frac{cg_k}{2})\cos[2\pi f_k t_d(R) + cp_k] \\ &\quad - j(1 + \frac{cg_k}{2})\sin[2\pi f_k t_d(R) + cp_k] \end{aligned} \tag{4.36}$$

from which, it is seen that the I and Q components become orthogonal in phase and balanced in amplitude; hence the complex vector $I(R) + jQ(R)$ starts rotating circularly during the movement of the metal plate at a fixed frequency. The magnitude of the rotating vector is then measured and stored. This procedure is repeated at each frequency of the RF operating frequency range. These measured magnitudes are used as reference data to compensate for the common amplitude errors.

After compensating for the common amplitude errors, the normalized complex vectors $I + jQ$ can be expressed using (4.36) as

$$\begin{aligned} I(R) + jQ(R) &= \cos[2\pi f_k t_d(R) + cp_k] \\ &\quad - j\sin[2\pi f_k t_d(R) + cp_k] \end{aligned} \tag{4.37}$$

From (4.37), the phase of the complex vector $I + jQ$ is obtained as

$$\phi(f_i) = 2\pi f_k t_d + 2\pi f_k \alpha + \beta_k \tag{4.38}$$

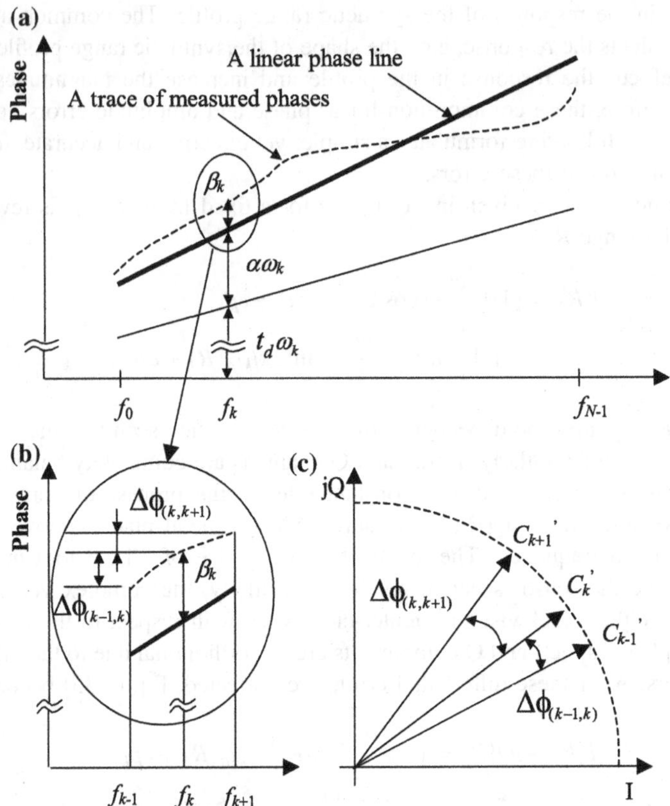

Fig. 4.17 Phase of the complex vector $I + jQ$ versus frequency: **a** linear transformation of the trace of calculated phases to a linear phase line, $(\alpha + t_d)\omega_k$, **b** a magnified drawing of (**a**) showing the trace of the calculated phases obtained by cumulating the phase differences $\Delta\phi_{0,1}$, …, $\Delta\phi_{k-1,k}$, …, $\Delta\phi_{N-2,N-1}$, **c** non-linearity of the calculated phases in polar form, where C_k' is the kth complex vector after compensating for the common amplitude deviation

with the aid of (4.34) and (4.35). As mentioned earlier, the non-linear phase error β_k needs to be corrected. Figure 4.17 illustrates the calculated phases $\phi(f_i)$ versus frequency. Cumulating the phase difference between two consecutive RF frequencies

$$\Delta\phi_{k-1,k} = 2\pi f_k t_d + 2\pi f_k \alpha + \beta_k - (2\pi f_{k-1} t_d + 2\pi f_{k-1}\alpha + \beta_{k-1}) \tag{4.39}$$

unwraps the calculated phases and makes it easy to draw the trace of the calculated phases as shown in Fig. 4.17a, b. Figure 4.17c shows that the rotation of the vector $I + jQ$ is not constant, which is due to the non-linear phase error β_k.

After drawing an appropriate linear phase line as shown in Fig. 4.17a, the non-linear phase error β_k is then determined by subtracting the linear phase line from the trace of the calculated phases. Consequently, the complex vector is obtained, after correcting the non-linear phase error, as

$$\begin{aligned} I(R)+jQ(R) &= \cos[2\pi f_k t_d(R)+2\pi f_k \alpha] - j\sin[2\pi f_k t_d(R)+2\pi f_k \alpha] \\ &= \exp\{-2\pi f_k[t_d(R)-\alpha]\} \end{aligned} \tag{4.40}$$

The non-linear phase error β_k at all the RF frequencies for the metal plate is stored in memory and used as reference data for compensating for the non-linear phase error of an actual target. The flow chart in Fig. 4.18 shows the procedure for extracting the common amplitude and non-linear phase errors. Figure 4.19 shows the common amplitude deviations and non-linear phase errors of the measured vectors for the metal plate used for correction in the frequency band of interest.

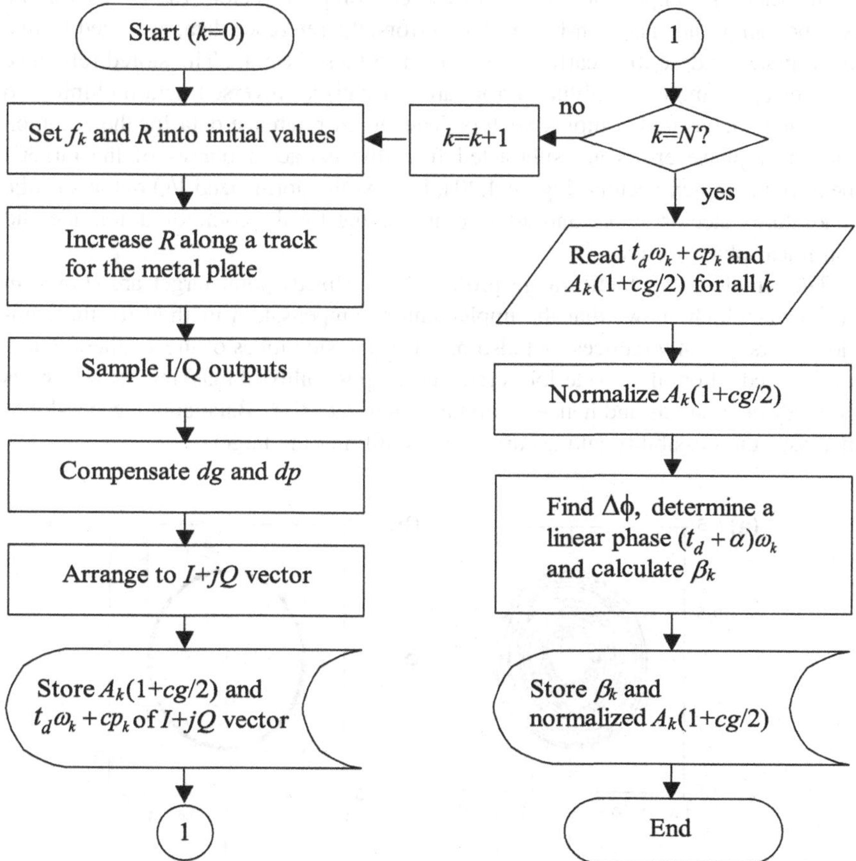

Fig. 4.18 Flow chart for calculating the common errors

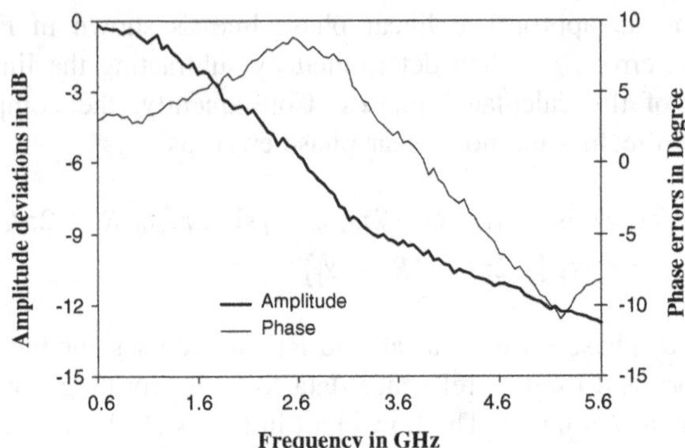

Fig. 4.19 Amplitude deviations and non-linear phase errors of the complex vectors due to the imperfection of the system

In order to compensate for the measured complex vectors of targets for the common amplitude and non-linear phase errors, the reference data, extracted from a metal plate, as described earlier, are applied to these vectors. The stored reference data for the common amplitude errors are normalized, inversed, and multiplied to the target's measured complex vectors. The stored reference data for the common non-linear phase errors are subtracted from the extracted phases of the target's measured complex vectors. Figure 4.20a, b show the normalized I/Q outputs of the quadrature detector before and after compensating for the common amplitude and non-linear phase errors.

The simulated synthetic range profiles for a (fixed) point target are shown in Fig. 4.21, which shows that the implemented compensation method for the common errors not only reduces, but also balances the side-lobes of the synthetic range profile. Reduction of the side-lobes reduces the possibility of masking the responses from adjacent targets and hence facilitating their detection. Balancing the side-lobes increases the possibility and accuracy in identifying the target.

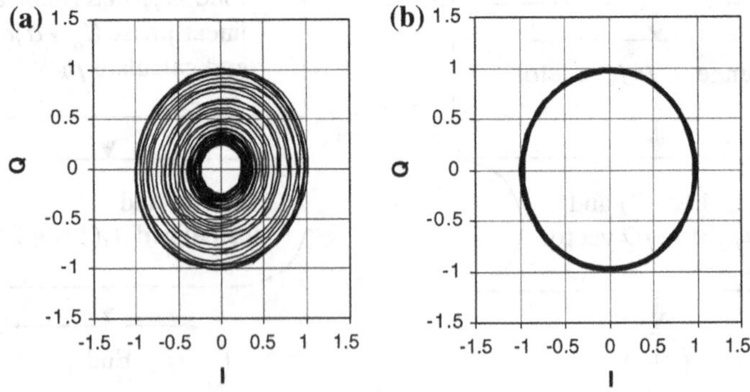

Fig. 4.20 Normalized I/Q (**a**) before and (**b**) after compensating for the amplitude deviations and non-linear phase errors of the quadrature detector

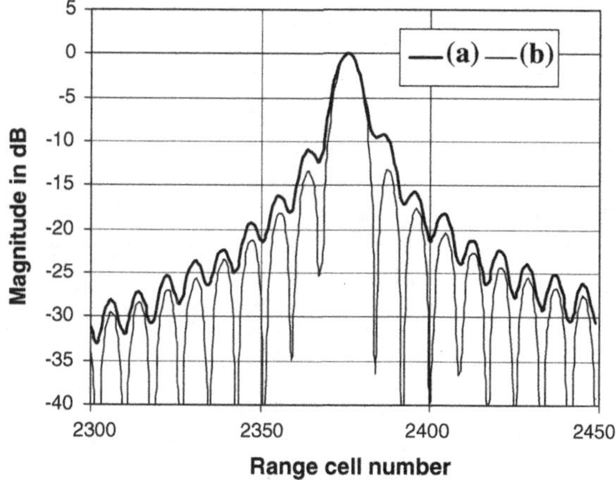

Fig. 4.21 Synthetic range profiles obtained from a fixed target (**a**) before and (**b**) after compensating for amplitude deviations and non-linear phase errors. The main peaks of the pulses indicate the target location

Upon compensating for the errors in the I and Q channels, the digital I and Q components are combined into a complex vector for each frequency step. An array V consisting of N complex vectors corresponding to N frequency steps is then formed as

$$V = [C_0, C_1, \ldots, C_{N-1}] \tag{4.41}$$

where $C_{N-1} = I_{N-1} + jQ_{N-1}$. Adding ($M$-$N$) zeros to the complex vector array V generates a new array V_M of M elements. This zero-padding is needed to improve the range accuracy, as discussed earlier, and the speed of the IDFT using Fast Fourier Transform (FFT). Finally, FFT is applied on the array V_M to get the output response's synthetic pulse. It is noted that an appropriate window function is also used to reduce the side-lobes of the synthetic range profile, which might mask other profiles produced by multiple targets. This window function can be selected based on individual target responses.

4.3 Summary

This chapter presents the development of two SFCW radar sensors realized in single packages with microwave integrated circuits (MICs) and microwave monolithic integrated circuits (MMICs): a millimeter-wave sensor operating across 29.72–37.7 GHz and a microwave sensor working from 0.6–5.6 GHz. The designs of the transceivers, antennas and signal processing, which form the most important parts of the sensors, are described. The presented information, although concise, is essential and in sufficient details, which enable microwave engineers to build similar microwave and millimeter-wave SFCW radar sensors for various sensing applications.

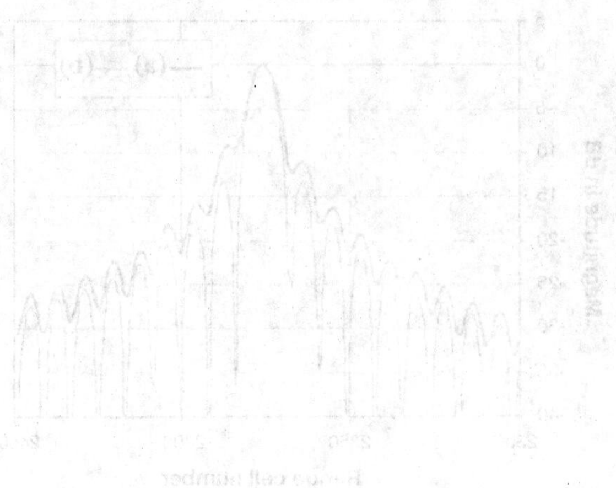

Chapter 5
Characterizations and Tests of Stepped-Frequency Continuous-Wave Radar Sensors

5.1 Introduction

One of the key requirements in system development is demonstrating the workability of the system for applications in practical settings. To that end, the designed microwave and millimeter-wave SFCW radar sensors described in Chap. 4 were used to conduct various laboratory and field tests. The electrical performances of these sensors were first evaluated in laboratory to verify their electrical parameters against the designed values such as the systems' dynamic ranges and transmitting powers. They were afterwards used to perform various measurements for surface and subsurface sensing applications. The systems' electrical performances and the results of some of the sensing tests including measurements of surface profile, liquid level and buried objects, and characterizations of pavement samples and actual roads are described in this chapter. The conducted sensing experiments discussed in this chapter give insightful overviews of the measurement procedures as well as the operations and performance-evaluations of these sensors for sensing applications.

5.2 Electrical Characterizations of Developed Stepped-Frequency Radar Sensors

The developed microwave and millimeter-wave SFCW radar sensors were first tested for their electrical performances. These tests are electrical tests making sure the system function properly as designed and are described in this section. The sensing tests for different applications are given in Sects. 5.3 and 5.4.

C. Nguyen and J. Park, *Stepped-Frequency Radar Sensors*,
SpringerBriefs in Electrical and Computer Engineering,
DOI 10.1007/978-3-319-12271-7_5

5.2.1 Microwave SFCW Radar Sensor

Figure 5.1 shows the block diagram of the microwave SFCW radar sensor. It consists of a transceiver, two antennae, and a signal processing block. Figure 5.2 shows the measured transmission gain G_T of the high frequency circuits' block of the transmitter (TX) between the IF input and the TX output ports over the entire operating frequency.

The output power at each component of the low frequency circuits of the transmitter was measured after adjusting the oscillator power with two attenuators to the desired specifications of −1 dBm at the splitter output, which was fed to the LO amplifier, and −4 dBm at the IF output. Table 5.1 shows the measured output power at each component of the transmitter. The measured transmitter output was in the range of 7.5–11.5 dBm.

The transmission gain of the high frequency circuit block of the receiver between the Rx input port and IF output port was measured over the entire operating frequency. In this measurement, the Rx input was connected to the Tx output through a 25-dB attenuator simulating the 25-dB transmission loss L_t mentioned in Chap. 4. Figure 5.3 shows the measured transmission gain.

Figure 5.4 shows the measured output power of the high frequency circuit block of the receiver at 3 GHz. The measured input 1-dB compression point of the high frequency circuit block was −4 dBm at 3 GHz. With the aid of a video amplifier that increases the maximum output power (−7.4 dBm) of the quadrature detector, the ADC's maximum input range constraint of 9 dBm was satisfied. Table 5.2 summarizes the receiver's measured electrical characteristics.

The system's dynamic range was measured at 3 GHz using a set-up as shown in Fig. 5.5, where an external attenuator was used to decrease the receiver's input power level. The base-band I/Q signals were then sampled and transformed to a synthetic pulse. The attenuation level was increased from 25 dB upwards, which

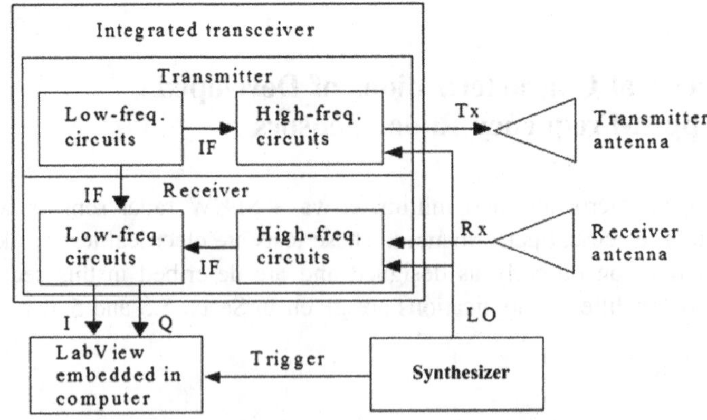

Fig. 5.1 Block diagram of the microwave SFCW radar sensor

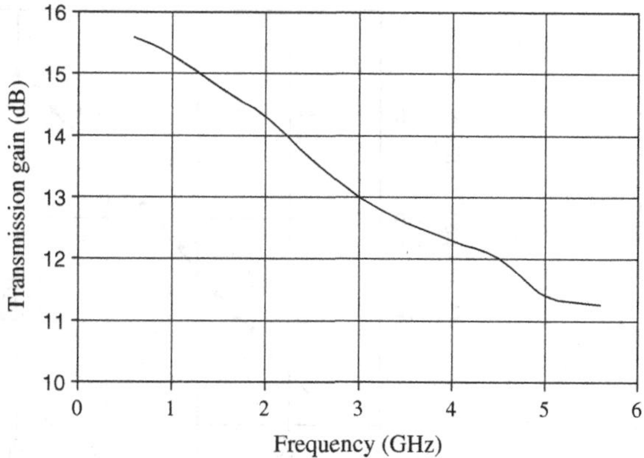

Fig. 5.2 Measured transmission gain of the high frequency circuits' block of the transmitter

Table 5.1 Measured output power of the transmitter

	Gain	Loss (−dB)	Pout (dBm)	G_T (dB)
STALO			4	
Attenuator		1	3	−1
LPF		0.5	2.5	−0.5
Splitter		3.2	−1	−3.5
Attenuator		3	−4	−1
Up-converter		N/A	N/A	11.4–15.5
1st Amplifier	N/A		N/A	
2nd Amplifier	N/A		7.4–11.5	
Total	N/A	N/A	–	5.4–9.5

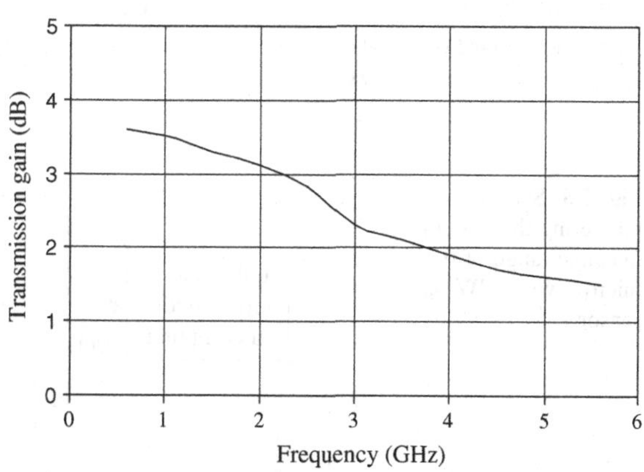

Fig. 5.3 Measured transmission gain of the high frequency circuit's block of the receiver

accounts for the transmission loss, until the synthetic pulse was below 10-dB SNR, which occurred at a level of 105 dB. Therefore, the actual system performance factor and the system's dynamic range were found to be 105 and 80 dB, respectively. The measured system dynamic range was slightly lower than the expected

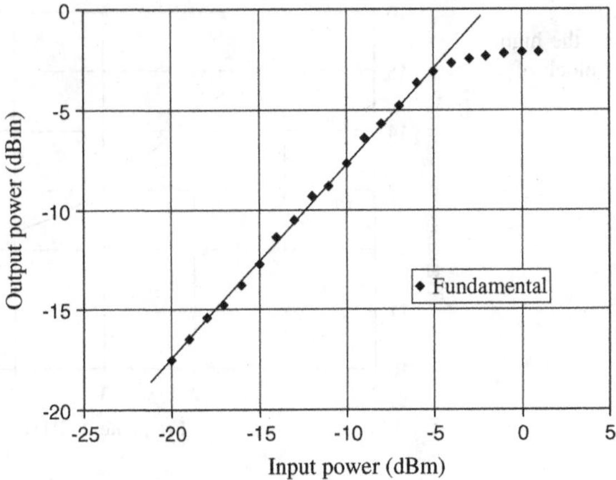

Fig. 5.4 Measured output versus input power of the receiver's high frequency circuit block at 3 GHz

Table 5.2 Measured output power at each component of the receiver where the input power was in the range of −17.6 to −13.5 dBm

	Gain (Vo/Vi) (dB)	Loss (−dB)	Pout (dBm)	G_T (dB)
LNA	N/A	N/A	N/A	1.5–3.6
Down-conv.		N/A	−16.1 to −9.9	
LPF		0.5	−16.6 to −10.4	–0.5
Amplifier	13		−4.6 to 1.6	12
I/Q mixer		8.5	−13.6 to −7.4	–9
LPF (Ro = 200 Ω)		0.2	−19.8 to −13.6	–6.2
Amp. (Ro = 1 KΩ)	29.6		2.8 to 9	22.6
Total	N/A	N/A	–	20.4–22.5

Fig. 5.5 Set-up for measuring the system's dynamic range of the microwave SFCW radar sensor

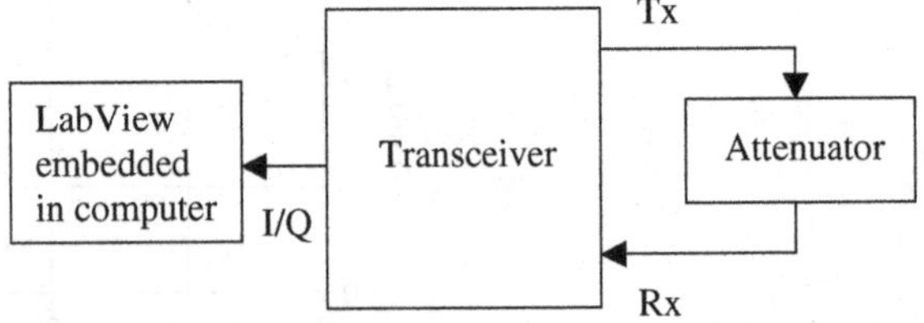

system dynamic range of 86 dB. The difference was mainly caused by the total gain deviation of around 5 dB at 3 GHz between the designed and measured gains. Table 5.3 shows other measured electrical characteristics and control parameters of the microwave SFCW radar sensor.

Table 5.3 Other measured electrical characteristics and control parameters of the microwave SFCW radar sensor

Electrical characteristics		Control parameters	
DR_s	80 dB	Freq. step Δf	10 MHz
SF_a (at 3 GHz)	105 dB	Number of freq. steps	500
DC Power consumption	2.9 W	*PRI*	100 ms
P_{1dB}	−4 dBm	ADC sampling freq.	1 kHz

5.2.2 Millimeter-Wave SFCW Radar Sensor

The millimeter-wave SFCW radar sensor employs a similar configuration as that of the microwave SFCW radar shown in Fig. 5.1, but it uses one antenna for both transmission and reception. Figure 5.6 shows the transmission gain G_T of the high frequency circuit block of the transmitter measured between the IF input and TX output ports from 27 to 36 GHz.

The output power of each component of the low frequency circuit portion of the transmitter was measured after the two attenuators were tuned to adjust the measured PLL oscillator power of 5.5 dBm to achieve the desired specifications 0 dBm at the splitter output fed to the LO amplifier and −0.5 dBm at the IF output. Table 5.4 shows the measured output power at each component of the transmitter where the measured transmitter output was in the range of 3.8–5.3 dBm.

Figure 5.7 shows the measured transmission gain of the high frequency circuit block of the receiver between the Rx input and IF output ports from 27 to 36 GHz.

Figure 5.8 shows the output power of the high frequency circuit block of the receiver at 32 GHz. The measured input 1 dB compression point of the high frequency circuit block was −6 dBm at 32 GHz.

The video amplifier was used to increase the maximum output power of −10 dBm of the quadrature detector to meet the ADC's maximum input range of 9 dBm. Table 5.5 shows the measured receiver's electrical characteristics.

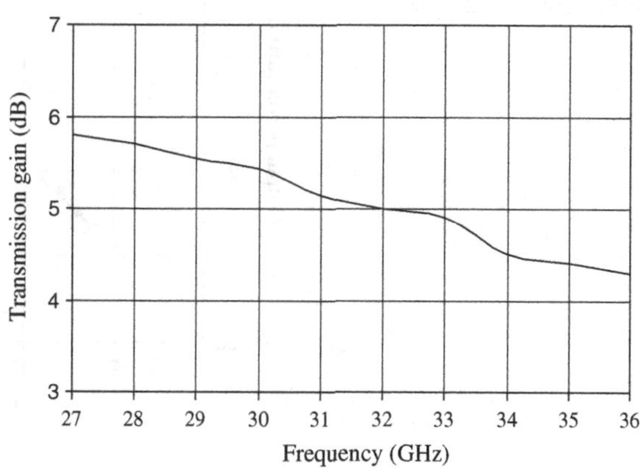

Fig. 5.6 Measured transmission gain of the high frequency circuit's block of the transmitter

Table 5.4 Measured output power at each component of the transmitter

	Gain	Loss (−dB)	Pout (dBm)	G_T (dB)
PLL osc.			4.5	
Attenuator		0	4	−0.5
LPF		0.5	3.5	−0.5
Spliter		3.5	0	−3.5
Attenuator		0	−0.5	−1.5
Up-converter		N/A	N/A	4.3−5.8
1st Amplifier	N/A		N/A	
2nd Amplifier	N/A		3.8–5.3	
Total	N/A	N/A	–	−1.7 to −0.2

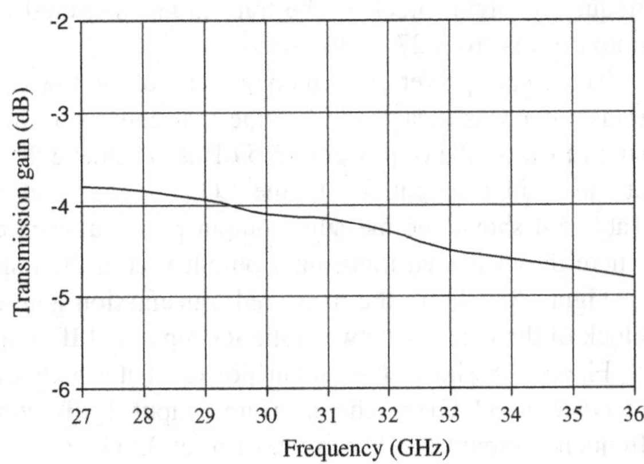

Fig. 5.7 Measured transmission gain of the high frequency circuit's block of the receiver

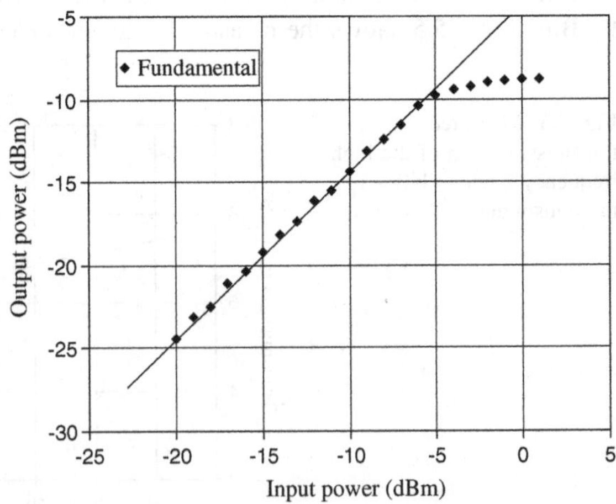

Fig. 5.8 Measured output versus input power of the high frequency circuit's block of the receiver at 32 GHz

Table 5.5 Measured output power of the receiver where the input power was in the range of −9.2 to −7.7 dBm

	Gain (Vo/Vi) (dB)	Loss (−dB)	Pout (dBm)	G_T (dB)
LNA	N/A	N/A	N/A	−4.8 to −3.8
Down-conv.		N/A	−14 to −11.5	
LPF		0.5	−14.5 to −12	−0.5
Amplifier	12		−3.5 to −1	11
I/Q mixer		8.5	−12.5 to −10	−9
LPF(Ro = 200)		0.2	−18.7 to −16.2	−6.2
Amp. (Ro = 1 k)	32.2		6.5 to 9	25.2
Total	N/A	N/A	–	15.7 to 16.7

Table 5.6 Other measured electrical characteristics and the control parameters of the system

Electrical characteristics		Control parameters	
DR_s	76 dB	Freq. step Δf	10 MHz
SF_a (at 32 GHz)	89 dB	Number of freq. steps	400
DC Power consumption	2.6 W	*PRI*	100 ms
P_{1dB}	−6 dBm	ADC sampling freq.	1 kHz

The system's dynamic range was not measured, as an external attenuator with a wide range of attenuation was not available. However, it was assumed to be 76 dB from the measured dynamic range value of the microwave SFCW radar sensor described earlier, taking into account the facts that the process gain of the millimeter-wave SFCW radar sensor is 1 dB lower than that of the microwave SFCW radar sensor and the gain deviation is 3 dB worse than the corresponding values of the microwave SFCW radar sensor. From the assumed value of the system's dynamic range, the actual system performance factor was calculated as 89 dB (76 + 13 dB). Table 5.6 shows other measured electrical characteristic and control parameters.

5.3 Tests of the Millimeter-Wave SFCW Radar Sensor

This section describes four different tests performed by the millimeter-wave SFCW radar sensor for sensing applications: surface profile measurement, liquid level measurement, and measurement of buried objects. In these tests, several assumptions were made. Firstly, the measured targets or objects were homogeneous materials with relatively low loss. Secondly, the incident waves were uniform plane waves. Thirdly, the double reflected waves in a layer were ignored. Lastly, the loss of the dry sand used in the burial mine sensing at Ka-band was simply estimated from those at 0.1 and 1 GHz presented in [31]. These assumptions may cause discrepancies between the actual penetration depths and the measured ones. However, the main purpose of these tests was to verify the feasibility of the millimeter-wave SFCW radar sensor for subsurface sensing and, hence, further investigation in this direction was not pursued. Nevertheless, the measured results agree well with the actual ones.

5.3.1 Measurement of Surface Profiles

This measurement demonstrates the ability of the millimeter-wave SFCW radar sensor in profiling the surfaces of structures. A plastic sample of 20 in. × 6 in. having its surface abruptly changed in height was used as an example for measuring the surface profile. The sensor's antenna was pointed directly onto the surface of the sample without contact while the sample was moving in the x direction as shown in Fig. 5.9. The measured data were collected every 0.2 in. along the direction of the movement. Figure 5.10 shows the reconstructed profile, obtained from the measurements, of the sample along with its actual profile. The reconstructed profile agrees well with the actual profile of the sample with less than ±0.04-in. error in height except near the edges, at which the actual surface profile abruptly changes.

In order to find the sensor's lateral resolution, we estimated a minimum cross-range that can be detected with this sensor from the reconstructed profile. The minimum cross-range is given by subtracting B from A, as shown in Fig. 5.10, as the distance of B-A is constant at a particular height, which means that the sensor can reconstruct the bottom surface when B is greater than 0. The estimated minimum detectable cross-range, R_{c_min}, representing the sensor's lateral resolution is 1 in., which agrees well with the lateral resolution of 0.92 in. obtained theoretically for the range R of 3.5 in. with the aid of Eq. (3.22) in Chap. 3.

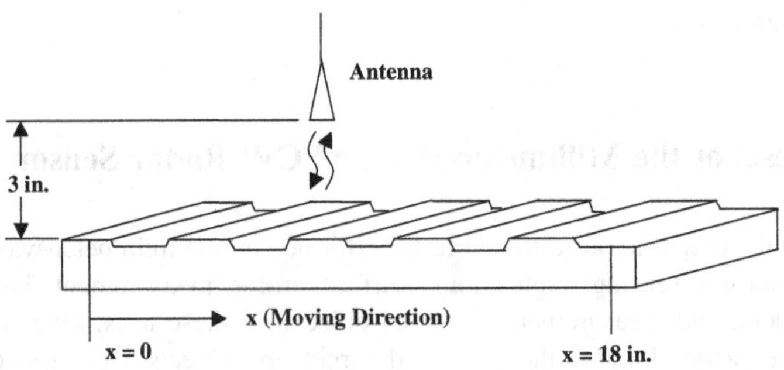

Fig. 5.9 Set up for measuring surface profile

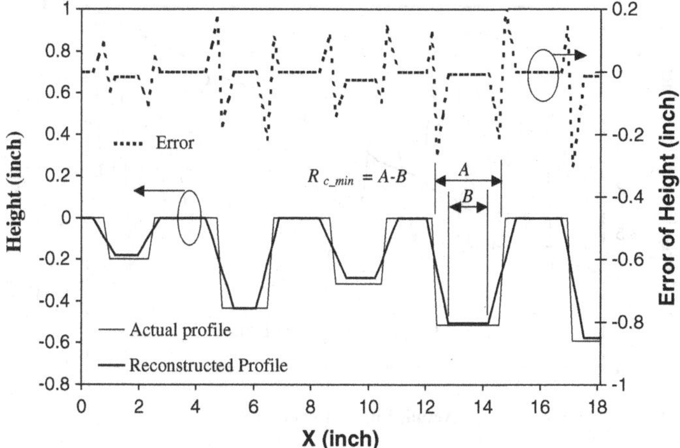

Fig. 5.10 Reconstructed and actual profiles of the surface of the sample in Fig. 5.9, where the height is set to 0 at the top surface at $x = 0$

5.3.2 *Measurement of Liquid Levels*

This measurement demonstrates the ability of the millimeter-wave SFCW radar sensor in monitoring continuously the liquid levels in storage tanks. Figure 5.11 illustrates the measurement of a liquid level in a tank by the sensor. The level of the liquid was decreased from the reference level, initially set at 0, to 3 in. below the reference and the resulting changes were measured. Figure 5.12 shows the measured liquid level against the actual one, demonstrating an excellent agreement between them with less than ±0.04-in. error.

According to the foregoing measurements of surface profiling and liquid level, the millimeter-wave SFCW radar sensor achieves a good range-accuracy with less than ±0.04-in. error, which agrees very well with the theoretical error of ±0.036 in., obtained from Eq. (3.17) in Chap. 3, where the frequency step and the number of

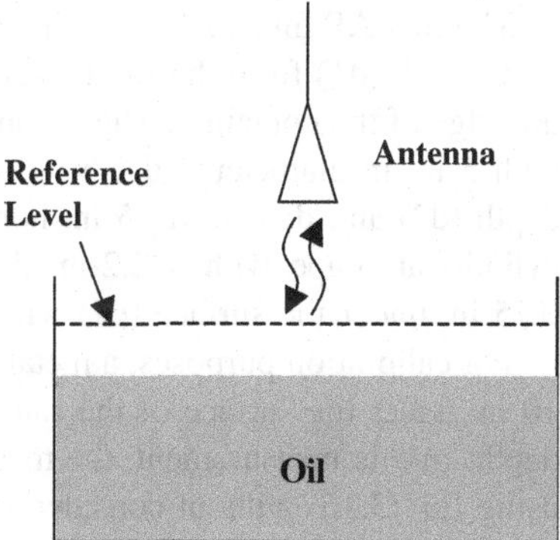

Fig. 5.11 Set up for measuring the liquid level in a tank

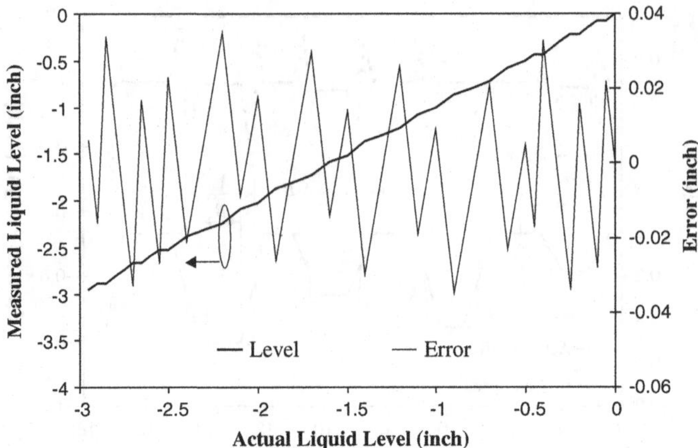

Fig. 5.12 Measured versus actual liquid levels in the tank in Fig. 5.11. The negative sign (−) means below the reference level

steps are 20 MHz and 4096 points, respectively. Also, a lateral resolution of 1 in. was obtained, which agrees quite well with the theoretical lateral resolution of 0. 72–0.92 in. In addition, the sensor provides very accurate measurement of the vertical displacement of liquid levels with less than ±0.04-in. error.

5.3.3 Measurement of Buried Objects

In this measurement, the millimeter-wave SFCW radar sensor shows its ability in sensing buried objects such as mines, unexploded ordnance (UXO), underground and structures, and pipes. To demonstrate possible usages for these subsurface sensing applications, three different anti-personnel (AP) metal mines buried in sand were used as shown in Fig. 5.13. For the sake of demonstration without elaborate image processing, the measurement was aimed only to localize and detect these AP mines buried under a sandy surface.

The first AP mine, AP1, is a metal sphere of 2.5 in. in diameter. It was buried at 2-in. depth (d1) from the sand's surface and displaced 7 in. horizontally (h1) from the edge of the container. The second AP mine, named AP2, is cylindrical in shape with 5 in. in diameter and 2.5 in. in height. This AP mine was buried at 6 in. of depth (d2) and displaced 15 in. horizontally (h2). The third AP mine, AP3, has a cylindrical shape with a 2.2-in. diameter and 3.5-in. height, and was buried at 0.75 in. under the surface (d2) with 23 in. in horizontal displacement (h3).

For calibration purposes, a metal plate having an area of 0.04 m^2 was first placed 10 in. under the surface of the sand and measurements were performed to find its depth. In this measurement, the measured depth of the metal plate was determined using Eq. (3.16) without considering the effect of the propagation medium; that is,

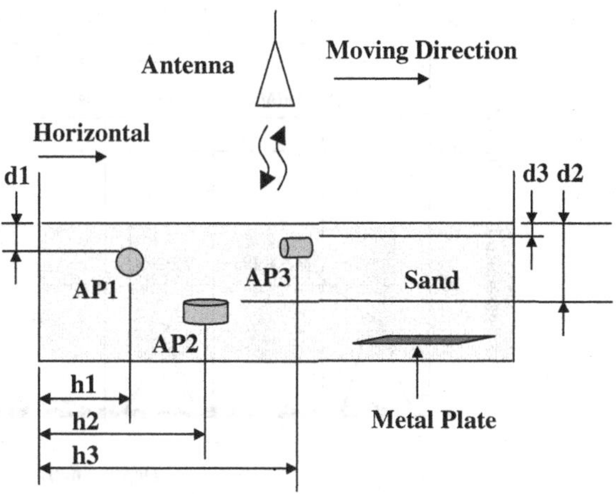

Fig. 5.13 Set up for detecting AP mines buried in sand

air is assumed for the medium. The measured depth of the metal plate was found to be 16.7 in., which is longer than the actual depth as the dielectric medium was not considered. From the measured depth d_m of the metal plate, the measured depth d in sand can be found using the following proportional expression

$$d = \frac{10d_m}{16.7} \tag{5.1}$$

Following the measurement for the metal plate, the depths (d1, d2, and d3) and horizontal displacements (h1, h2, and h3) of the AP mines were then measured with the antenna moving in the horizontal direction as shown in Fig. 5.13. Using Eq. (3.16), the measured depths of AP1, AP2, and AP3 mines, assuming air as the propagating medium, were recorded as 3.39, 10.09 and 1.33 in., respectively. Therefore, the measured depths of AP1, AP2, and AP3 mines in sand were found at 2.05, 6.08 and 0.8 in. as seen in Fig. 5.14, which shows the synthetic pulses corresponding to these AP mines and the metal plate. The measured horizontal displacements of AP1, AP2, and AP3 mines were 7, 15.75, and 23.25 in., respectively. Figure 5.15 shows the detected and localized mines. The results show that the measured horizontal displacements and depths of the AP mines are fairly close to the corresponding actual horizontal displacements and depths with less than 0.75 and 0.08 in. of error.

Note that the AP3 mine buried at only 0.75 in. under the top surface was clearly detected and localized, which conforms the theoretical range resolution of 0.59 in. calculated using Eq. (3.27). As a result, the sensor demonstrated its ability in detecting and locating burial objects of very small sizes with a competitively high resolution.

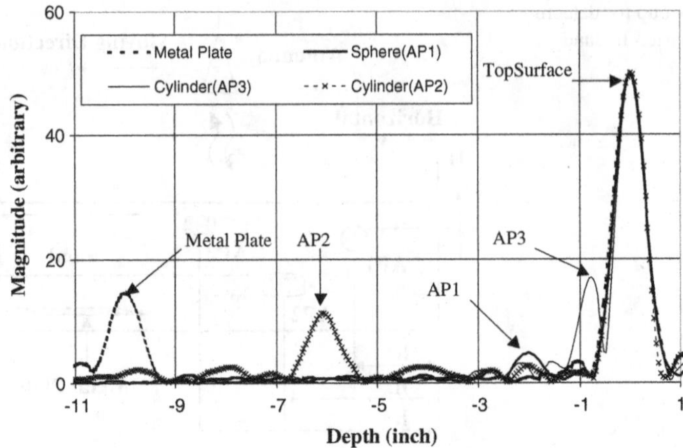

Fig. 5.14 Synthetic pulses extracted from the measurements of the AP mines and metal plate

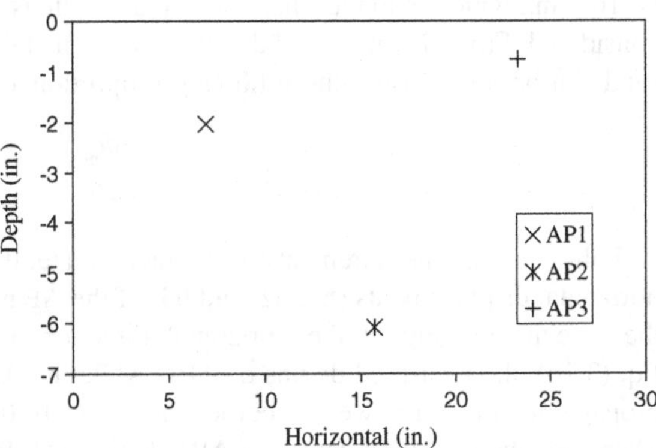

Fig. 5.15 AP mines localized in depth and horizontal displacement

5.4 Tests of the Microwave SFCW Radar Sensor

In this section, two measurements performed using the microwave SFCW radar sensor to verify its feasibility for subsurface sensing are described: one conducted on a pavement sample in laboratory and the other done on actual roads in a test site at the Texas A&M University. Several assumptions were made in these measurements. Firstly, it was assumed that the targets or objects were homogeneous materials with low loss. Secondly, the incident waves were assumed as uniform plane waves. Thirdly, multiple-reflected waves in a layer were ignored as mentioned in Chap. 2. Lastly, the layers were taken to be smooth half-spaces. These assumptions are not correct for practical pavement materials. However, as will be seen later, accurate measured results were achieved for practical pavement structures.

5.4.1 Measurement of Pavement Sample

This measurement demonstrates the ability of the microwave SFCW radar sensor in characterizing multi-layer structures. A pavement sample constructed with two layers in a wooden box of 36 in. × 36 in., as shown in Fig. 5.16, was used for the measurement. The top layer is asphalt having a thickness of 2.6–2.7 in. while the bottom layer is the base with a thickness of 4.1 in. and filled with limestone. The sensor's two antennae, located at 0.2 m of stand-off distance from the sample, were pointed obliquely onto the sample's surface with a parallel polarization and 10° of incident angle through the air. The sensor was operated at 3 GHz.

As shown in Chaps. 2 and 3, an incident angle of 10° is not large and has little effects on the reflection and transmission coefficients as well as the penetration depth; therefore it was ignored. The signals reflected at the interfaces between the sample's layers are illustrated in Fig. 5.16.

The reflected electric field E_{r2} at the 2nd interface in Fig. 5.16 is obtained for normal incidence, using Eq. (2.34) in Chap. 2, as

$$E_{r2} = E_i \mathrm{T}_{10} \mathrm{T}_{01} \Gamma_{21} \exp(-\alpha_1 d_1) \tag{5.2}$$

where E_i represents the electric field of the incident wave.

Assuming the dielectric layers have low loss and hence their relative dielectric constant can be approximately considered as real, the relative dielectric constant $\varepsilon_{r(i+1)}$ of layer i + 1 (i = 0, 1, 2) can be derived from Eq. (2.31) for a uniform plane wave incident normally from layer i to layer i + 1 as

$$\varepsilon_{r(i+1)} = \varepsilon_{r(i)} \left(\frac{1 - \Gamma_{i+1i}}{1 + \Gamma_{i+1i}} \right)^2 \tag{5.3}$$

where $\varepsilon_{r(i)}$ is the relative dielectric constant of layer i and Γ_{i+1i} is the reflection coefficient of the wave incident from layer i to layer i + 1.

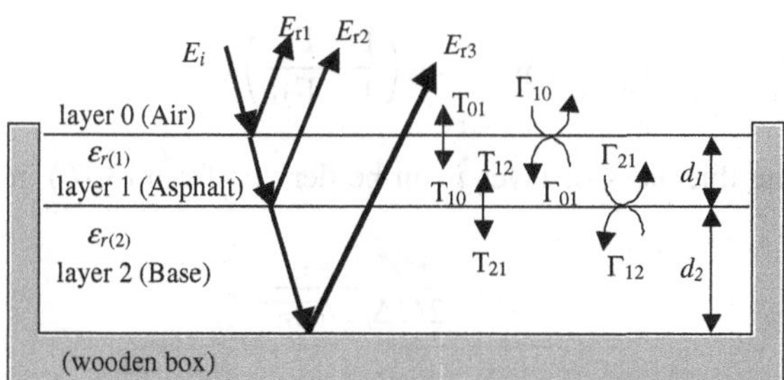

Fig. 5.16 Sketch of the pavement sample in a wooden box together with the incident and reflected waves. E_i is the incident wave; E_{r1}, E_{r2} and E_{r3} are the reflected waves at the interfaces between layers 0 and 1, layers 1 and 2, and layer 2 and the wooden box, respectively, and d_1 and d_2 are the thickness of layers 1 and 2, respectively

The magnitude of the reflected electric field E_{rn} from the nth interface can be found from Eq. (2.37) for normal incidence as

$$|E_{rn}| = |E_i||\Gamma_{nn-1}|\left(\prod_{m=1}^{n-1} \mathrm{T}_{mm-1}\mathrm{T}_{m-1m}\exp(-2\alpha_m d_m)\right) \tag{5.4}$$

where α_m and d_m are the attenuation constant and thickness of layer m.

The magnitude of the incident electric field E_i is equal to that of the electric field E_m reflected from a (perfect) metal plate placed at the first interface corresponding to $\Gamma_{10} = -1$. That is,

$$|E_i| = |E_m| \tag{5.5}$$

The reflection coefficients at the first and second interfaces can then be obtained from Eqs. (5.4) and (5.5) as

$$\Gamma_{10} = \frac{|E_{r1}|}{|E_m|} \tag{5.6}$$

and

$$\Gamma_{21} = \frac{|E_{r2}|}{|E_m|\mathrm{T}_{10}\mathrm{T}_{01}\exp(-2\alpha_1 d_1)} \tag{5.7}$$

respectively.

The relative dielectric constants of the asphalt and base layers can be obtained from (5.3), where $\varepsilon_{r(0)} = 1$ for the air incident medium, as

$$\varepsilon_{r(1)} = \left(\frac{1-\Gamma_{10}}{1+\Gamma_{10}}\right)^2 \tag{5.8}$$

and

$$\varepsilon_{r(2)} = \varepsilon_{r(1)}\left(\frac{1-\Gamma_{21}}{1+\Gamma_{21}}\right)^2 \tag{5.9}$$

respectively. The thickness of layer i can be derived from (3.16) in Chap. 3 as

$$d_i = \frac{v\Delta n_{(i,i+1)}}{2M\Delta f\sqrt{\varepsilon_{r(i)}}} \tag{5.10}$$

where $\Delta n_{(i,i+1)}$ is the difference between the bin numbers of cells corresponding to layers i and $i + 1$ as shown in Fig. 5.17, which shows the synthetic range profiles of the pavement sample and the metal plate obtained from the measurement data. It

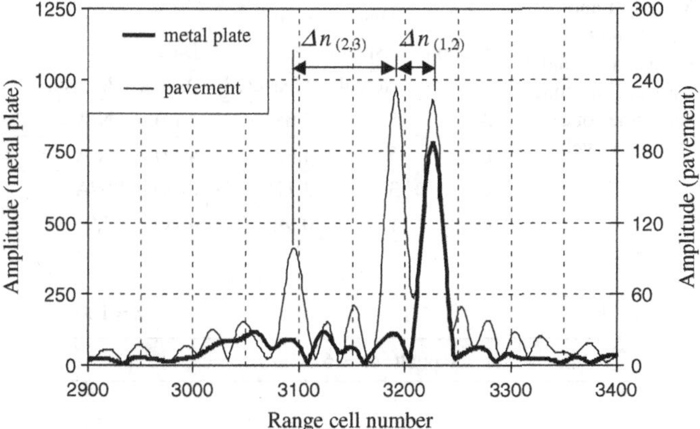

Fig. 5.17 Synthetic range profiles obtained from the pavement sample and metal plate

Table 5.7 Comparison between the actual and measured parameters of the pavement sample

Parameter		Asphalt	Base
Relative dielectric constant	Measured	3.24	12.5
Thickness (in.)	Actual	2.6–2.7	4.1
	Measured	2.72	4.04

should be mentioned here that this procedure can also be used for structures containing more than 3 layers.

Table 5.7 shows the measured parameters of the pavement sample along with the actual values. The measured thickness of each layer agrees well with the actual value. Note that the theoretical relative dielectric constants of the sample's asphalt and base materials listed in Table 2.1 of Chap. 2 are not used here since these values vary over a wide range and the asphalt and base materials in Table 2.1 are not the same as those in our pavement sample.

5.4.2 *Measurements of Actual Roads*

These measurements demonstrate the ability of the microwave SFCW radar sensor in characterizing multi-layer structures in practical settings. The measurements were conducted on actual roads at two test sites of the Texas A&M University. As discussed earlier, only stationary tests were performed due to the low *PRI* of the available synthesizer.

The first measurement was conducted on the test site that was partitioned with several different sections. These sections contain pavement layers with various thicknesses as shown in Table 5.8. However, the actual thickness and other properties should be different with those values in Table 5.8 as the test site was constructed around 1972. The second measurement was performed on a road where the

Table 5.8 Materials and thicknesses of different pavement sections. x, y, and z indicate limestone, limestone +2 % lime, and limestone +4 % cement, respectively

Section	Thickness (in.)				
	Asphalt		Base		Subgrade
	Actual	Measured	Actual	Measured	Actual
A	5	4.68	4 (z)	N/A	4 (x)
B	5	4.82	4 (z)	N/A	4 (z)
C	3	3.12	8 (y)	N/A	8 (y)
D	1	0.99	4 (y)	3.96	12 (x)

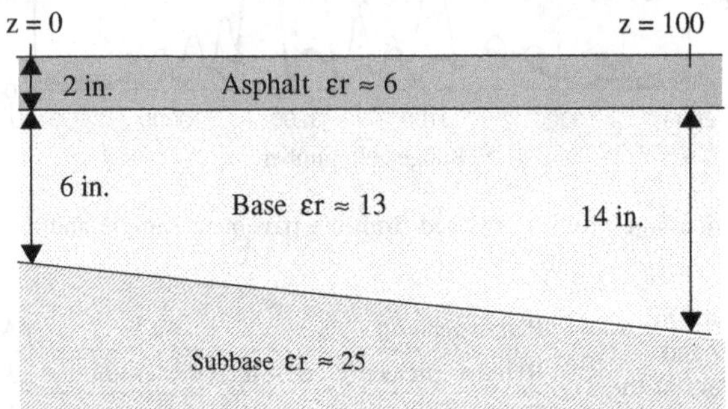

Fig. 5.18 Cross-section of the road

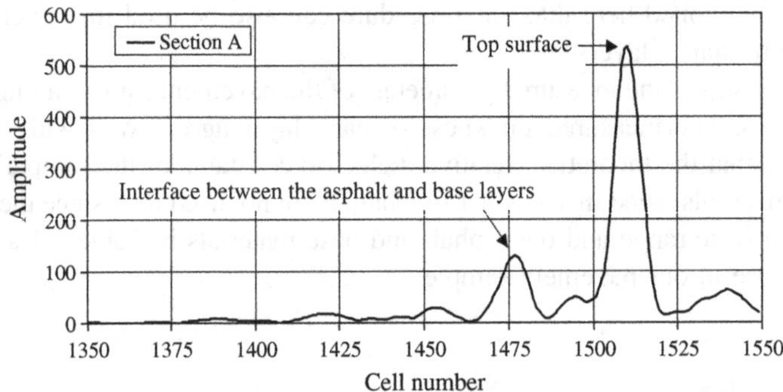

Fig. 5.19 Synthetic profile of section A

thickness of the asphalt layer is fixed at 2 in. while the thickness of the base layer is varied continuously as shown in Fig. 5.18. The measurement data were collected every 20 ft for a total length of 100 ft.

Figures 5.19, 5.20, 5.21, 5.22 show the synthetic profiles obtained from the measurements of sections A-D. As shown in these figures, the sensor can detect up to the second interface between the asphalt and base layers for sections A, B, and C and up to the third interface between the base and subgrade layers for section D, where the thickness of the asphalt layer is only 1 in. The measured thicknesses of the asphalt and base layers were found using (5.10) and shown in Table 5.8.

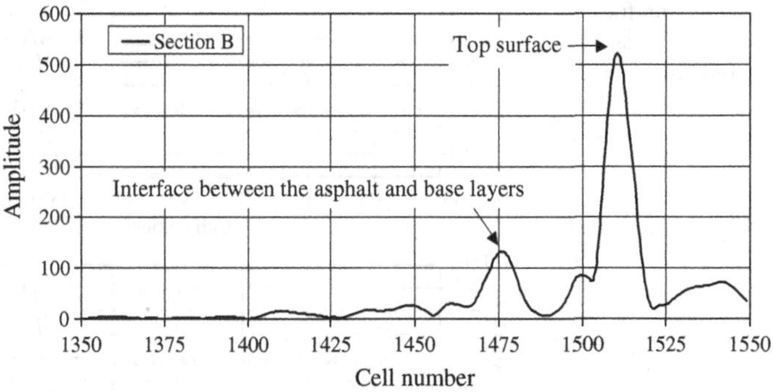

Fig. 5.20 Synthetic profile of section B

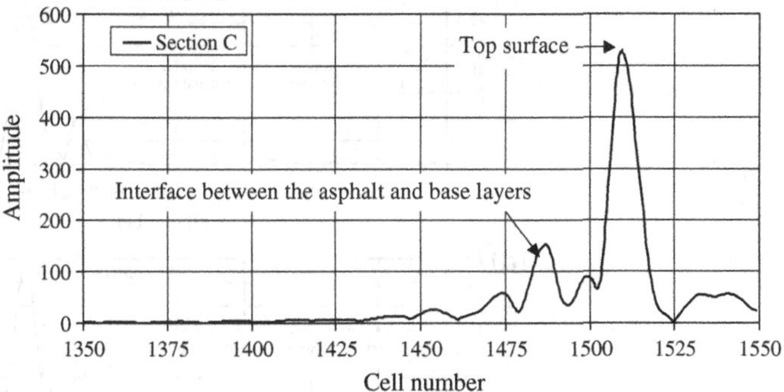

Fig. 5.21 Synthetic profile of section C

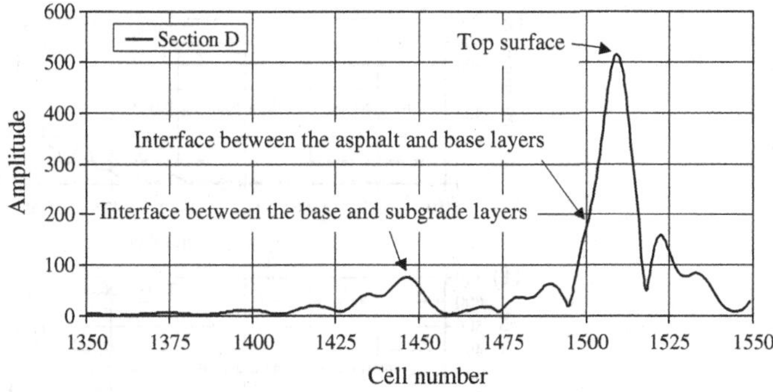

Fig. 5.22 Synthetic profile of section D

Figure 5.23a–f show the synthetic profiles obtained at different locations of the road shown in Fig. 5.18. As expected from the previous measurement results, the sensor can only detect the second interface between the asphalt and base layers.

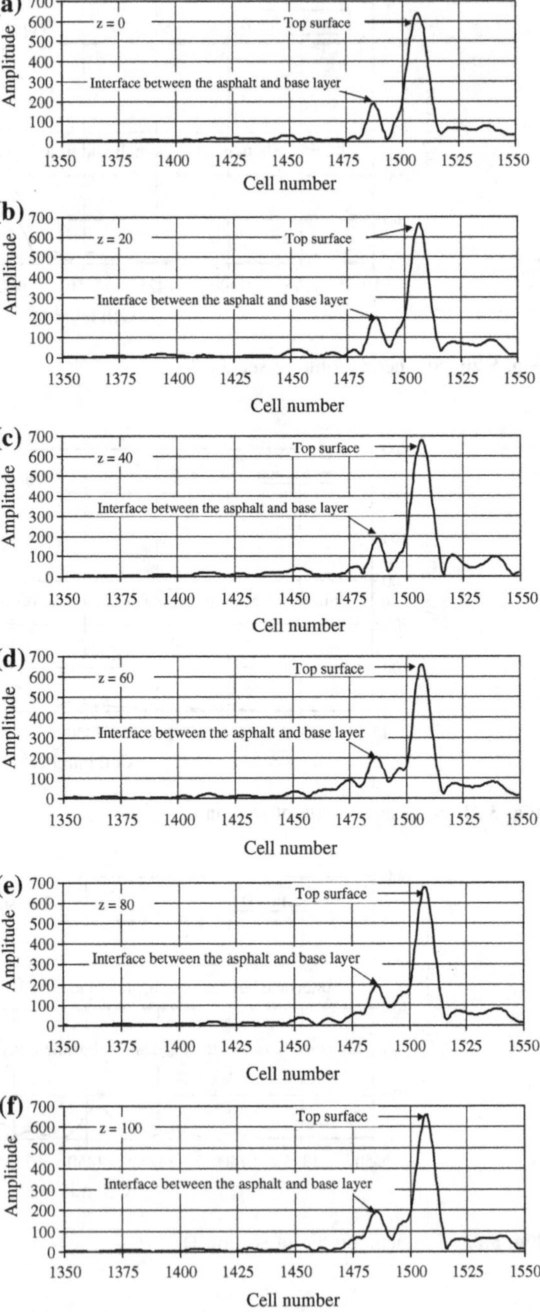

Fig. 5.23 Synthetic profiles of the road in Fig. 5.18 at different locations from z = 0 to 100 ft

Table 5.9 Comparison between the actual and measured thicknesses of the asphalt layer of the road shown in Fig. 5.18

Position z (ft)		0	20	40	60	80	100
Asphalt thickness (in.)	Actual	2	2	2	2	2	2
	Measured	2.13	2.13	2.13	2.25	2.25	2.25

Table 5.9 summarizes the measured and actual thicknesses of the asphalt layer, showing less than 0.25 in. of error between them.

From the measured results, it can be deduced that the sensor's transmitted power rapidly decreased in the layers due to significant loss incurred in these layers at high frequencies. However, these results demonstrate that the developed millimeter-wave SFCW radar sensor can clearly detect as least 5 in. of asphalt layers.

5.5 Summary

This chapter discusses the electrical tests and several sensing measurements of the developed microwave and millimeter-wave SFCW radar sensors described in Chap. 4. The electrical tests verify the electrical performances of these systems to make sure they possess the desired electrical characteristics as designed. These electrical tests, while do not directly show the actual performance of the sensors in practical sensing applications, are essential in system development and need to be conducted first. The sensing measurements of the millimeter-wave SFCW radar sensor described in this chapter include surface profiling, liquid-level measurement, and detection and location of buried objects. The sensing measurements for the microwave SFCW radar sensor are the measurements of pavement structures including an actual pavement in laboratory and actual roads. The measured sensing results show good agreement with the actual values, demonstrating the workability and usefulness of the developed sensors for various surface and subsurface sensing applications. Moreover, this chapter also describes of the procedures involved in the sensing measurements, which provide not only the operations of these sensors, but also their system evaluations for sensing applications.

Chapter 6
Summary and Conclusion

In the last five chapters, this book has presented the theory, analysis, and design of SFCW radar sensors and their components. Specifically, the book addresses the system analysis, transmitter design, receiver design, antenna design, signal processing, and system integration and test of two SFCW radar sensors. These systems are realized in a single unit using MICs and MMICs and operate from 29.72 to 37.7 GHz and 0.6 to 5.6 GHz. They can be used for surface and subsurface sensing applications and have demonstrated high resolution and accuracy in various sensing measurements.

Chapter 1 introduces the SFCW radar sensors and possible applications in sensing.

Chapter 2 addresses the general analysis of radar sensors, particularly signal propagation and scattering from objects, Friis transmission equation and radar equations, signal-to-noise ratio, receiver sensitivity, maximum range, and system performance factor. A set of modified radar equations that accurately characterize a subsurface radar sensor intended for sensing multi-layered structures, such as pavements, or to detect and localize buried objects under a surface, were derived. These equations involve more number of parameters than conventional equations and are needed in subsurface radar sensors in order to estimate accurately their penetration depths.

Chapter 3 covers the analysis of SFCW radar sensors. The sensors' operating principles and characteristics—from transmitted to received RF signals to down-converted I and Q signals and target's synthetic profile formed from the I and Q signals—are discussed. It also addresses the sensors' other important parameters such as angle and range resolutions, frequency step, number of frequency steps, total bandwidth, range, range accuracy and ambiguity, pulse repetition interval, dynamic range, system performance factor, and estimations of the maximum ranges in sensing multi-layer and buried targets.

C. Nguyen and J. Park, *Stepped-Frequency Radar Sensors*,
SpringerBriefs in Electrical and Computer Engineering,
DOI 10.1007/978-3-319-12271-7_6

In Chapter 4, the developments of the millimeter-wave SFCW radar sensor working across 29.72–37.7 GHz and the microwave SFCW radar sensor operating from 0.6 to 5.6 GHz are presented. They are based on a coherent super-heterodyne architecture and completely realized using MICs and MMICs in single units, which have low cost, light weight and small size. The design of the transceivers and antennas, the development of the signal processing, and the integration of these sensors are described. Specifically, a simple yet effective technique for compensating the common I/Q errors caused by the system itself is presented.

Chapter 5 covers the electrical tests and sensing measurements of the microwave and millimeter-wave SFCW radar sensors. The electrical tests validate the electrical performances of these systems. The measured system performance factors of the microwave and millimeter-wave SFCW radar sensors are 105 and 89 dB, respectively. The sensing measurements of the millimeter-wave SFCW radar sensor include surface profiling, liquid-level measurement, and detection and location of buried objects. The millimeter-wave SFCW radar sensor can profile the surface of a sample whose height rapidly changes along the horizontal direction with 1 inch of lateral resolution and less than ±0.04 inch of range accuracy. In addition, it accurately measured the displacement of liquid level with less than ±0.04 inch of discrepancy. It also demonstrates its benefits as a subsurface radar sensor in detecting and localizing very small buried AP mines under sand with less than 0.75 inch of vertical resolution. The sensing measurements for the microwave SFCW radar sensor are for pavements in laboratory and fields. The microwave SFCW radar sensor demonstrates its excellent performance with good measurement results on the laboratory's sample pavement with less than ±0.1 inch of error. It also shows that the thickness of the asphalt layer on the actual roads can be accurately measured with less than 0.25 inch of error. These measured sensing results show good agreement between the measured and actual data, demonstrating the workability, good performance, and usefulness of the developed sensors for various surface and subsurface sensing applications.

SFCW radar sensors are attractive for surface and subsurface sensing in various applications from military to civilian and commerce. As stated in Chap. 1, there are many aspects in SFCW radar sensors and a complete coverage would require a book of substantial size, which is beyond the scope of this book. The primary purpose of this book is to address only the essential parts of SFCW radar sensors including system and component analysis, design, signal processing, and measurement in a concise manner, yet with sufficient details, to allow the readers to understand and successfully design SFCW radar sensors for their intended applications, whether for research or for commercial usage. The measurements in surface and subsurface sensing for these systems indeed demonstrate not only the systems' workability, but also their usefulness in various surface and subsurface sensing. It should be noted that only a simple signal processing was implemented in the developed microwave and millimeter-wave SFCW radar sensors. For advanced applications of the developed SFCW radar sensors as well as other SFCW systems, advanced signal processing is needed to improve the systems' performance and capability and enhance their applications, such as increased lateral and range resolutions for better surface and subsurface sensing, compensation for the dispersion of propagation media to enhance the synthetic pulse's shape for further improvement of the range resolution, etc. For large-range sensing, high transmitting powers need to be used. For applications involving fast moving platforms like vehicles or aircrafts, fast frequency synthesizers need to be use in the transmitters.

Lastly, it is noted that the SFCW radar sensors presented in this book is not intended for complete field operations. Rather, they are simple laboratory prototypes used to demonstrate the design, analysis, and applications of SFCW systems. As such, the developed systems are not optimum, both physically and electrically. However, with the materials provided in this book, it should be relatively straightforward to design and produce field-ready SFCW systems for complete field operations with better manufacturing for more physically rigid systems and size reduction to make them more portable.

References

1. D.J. Daniels, *Surface Penetrating Radar* (IEE Press, London, U.K., 1996)
2. D.J. Daniels, D.J. Gunton, H.F. Scott, Introduction to subsurface radar. IEE Proc. **135**, 278–320 (1988)
3. S.L. Earp, E.S. Hughes, T.J. Elkins, R. Vikers, Ultra-wideband ground-penetrating radar for the detection of buried metallic mines. IEEE Aerosp. Electron. Syst. Mag. **11**, 30–39 (1996)
4. A. Langman, M.R. Inggs, A 1-2GHz SFCW radar for landmine detection, in *Proceedings of the 1998 South African Symposium*, Sept. 1998, pp. 453–454
5. C.J. Vaughan, Ground-penetrating radar surveys used in archaeological investigations. Geophysics **51**(3), 595–604 (1986)
6. J. Otto, Radar applications in level measurement, distance measurement and nondestructive material testing, in *Proceedings of the 27th European Microwave Conference and Exhibition*, vol. 2 (1997), pp 1113–1121
7. T. Lasri, B. Dujardin, Y. Leroy, Microwave sensor for moisture measurements in solid materials. Microwaves Antennas Propag. **138**, 481–483 (1991)
8. J.S. Park, C. Nguyen, A new millimeter-wave step-frequency radar sensor for distance measurement. IEEE Microwave Wireless Compon. Lett. **12**(6), 221–222 (2002)
9. J.S. Lee, C. Nguyen, T. Scullion, A novel compact, low-cost impulse ground penetrating radar for nondestructive evaluation of pavements. IEEE Trans Instrumen. Measur. **IM-53**(6), 1502–1509 (2004)
10. J.S. Park, C. Nguyen, An ultra-wideband microwave radar sensor for nondestructive evaluation of pavement subsurface. IEEE Sensors J. **5**, 942–949 (2005)
11. J.S. Park, C. Nguyen, Development of a new millimeter-wave integrated-circuit sensor for surface and subsurface sensing. IEEE Sensors J. **6**, 650–655 (2006)
12. J. Han, C. Nguyen, Development of a tunable multi-band UWB radar sensor and its applications to subsurface sensing. IEEE Sensors J. **7**(1), 51–58 (2007)
13. D.A. Ellerbruch, D.R. Belsher, Electromagnetic technique of measuring coal layer thickness. IEEE Trans. Geosci. Electron. **16**(2), 126–133 (1978)
14. E.K. Miller, *Time-domain Measurements in Electromagnetics* (Van Nostrand Reinhold Company, New York, NY, 1986)
15. C.H. Lee, Picosecond optics and microwave technology. IEEE Trans. Microwave Theory Tech. **38**, 569–607 (1990)
16. L.L. Molina, A. Mar, F.J. Zutavern, G.M. Loubriel, M.W.O'Malley, Sub-nanosecond avalanche transistor drivers for low impedance pulsed power applications, in *Pulsed Power Plasma Science-2001*, vol. 1 (2001), pp. 178–181
17. J.S. Lee, C. Nguyen, T. Scullion, New uniplanar subnano-second monocycle pulse generator and transformer for time-domain microwave applications. IEEE Trans. Microwave Theory Tech. **MTT-49**(6), 1126–1129 (2001)

C. Nguyen and J. Park, *Stepped-Frequency Radar Sensors*,
SpringerBriefs in Electrical and Computer Engineering,
DOI 10.1007/978-3-319-12271-7

18. J.W. Han, C. Nguyen, On the development of a compact sub-nanosecond tunable monocycle pulse transmitter for UWB applications. IEEE Trans. Microwave Theory Tech. **MTT-54**(1), pp 285–293 (2006)
19. P. Dennis, S.E Gibbs, Solid-state linear FM/CW radar systems-their promise and their problems, in *IEEE MTT-S International Microwave Symposium Digest*, vol. 74, no. 1 (1974), pp. 340–342
20. S.O. Piper, *Frequency-Modulated Continuous Wave Systems* (Norwood, MA, Artech House, 1993)
21. A.E. Carr, L.G. Cuthbert, A.D. Oliver, Digital signal processing for target detection in FMCW radar, in *IEE Proceedings of Communications, Radar, and Signal Processing*, vol. 128, no. 5 (1981), pp. 331–336
22. L.A. Robinson, W.B. Weir, L. Young, An RF time-domain reflectometer not in real time, in *GMTT International Microwave Symposium Digest*, vol. 72, no. 1 (1972), pp. 30–32
23. D.R. Wehner, *High Resolution Radar* (Norwood, MA, Artech House, 1995)
24. K. Iizuka, A.P. Freundorfer, Detection of nonmetallic buried objects by a step frequency radar. IEEE Proc. **71**(2), 276–279 (1983)
25. D.A. Noon, Stepped-frequency radar design and signal processing enhances ground penetrating radar performance, Ph.D. Thesis, University of Queensland, Queensland, Australia, 1996
26. R.C. Pippert, K. Soroushian and R.G. Plumb, Development of a ground-penetrating radar to detect excess moisture in pavement subgrade, in *Proceedings of the Second Government Workshop on GPR—Advanced Ground Penetrating Radar: Technologies and Applications,* Oct. 1993, pp. 283–297
27. A. Langman, P.D. Simon, M. Cherniakov, I.D. Langstaff, Development of a low cost SFCW ground penetrating radar, in *IEEE Geoscience and Remote Sensing Symposium*, vol. 4 (1996), pp. 2020–2022
28. G.F. Stickley, D.A. Noon, M. Cherniakov, I.D. Longstaff, Preliminary field results of an ultra-wideband (10–620 MHz) stepped-frequency ground penetrating radar, in *Proceedings of the 1997 IEEE International Geoscience and Remote Sensing Symposium*, vol. 3 (1997), pp. 1282–1284
29. D. Huston, J.O. Hu, K. Muser, W. Weedon, C. Adam, GIMA ground penetrating radar system for monitoring concrete bridge decks. J. Appl. Geophys. **43**, 139–146 (2000)
30. R.H. Church, W.E. Webb, J. B. Salsman, Dielectric properties of low-loss materials. Report of Investigations 9194, US Bureau of Mines, Washington D.C., 1998
31. J.L. Davis, A.P. Annan, Ground-penetrating radar for high-resolution mapping of soil and rock stratigraphy. Goephys. Prospect. **37**(5), 531–551 (1989)
32. M.I. Skolnik, *Introduction to RADAR Systems*, 3rd edn. (NY, McGraw-Hill, New York, 2001)
33. A.P. Annan, J.L. Davis, Radar range analysis for geological materials. Geol. Surv. Can. (77-1B), 117-124 (1977)
34. C. Nguyen, *Analysis Methods for RF, Microwave and Millimeter-Wave Planar Transmission Line Structures* (Wiley, New York, 2000)
35. S.T. Kim, C. Nguyen, On the development of a multifunction millimeter-wave sensor for displacement sensing and low-velocity measurement. IEEE Trans Microwave Theory Tech. **52**(11), 2503–2512 (2004)
36. C. Huynh, C. Nguyen, New ultra-high-isolation RF switch architecture and its use for a 10–38 GHz 0.18-μm BiCMOS Ultra-Wideband Switch. IEEE Trans. Microwave Theory Tech. **59**(2), 345–353 (2011)
37. J.G. Proakis, D.G. Manolakis, *Digital Signal Processing* (Englewood Cliffs, NJ, Prentice Hall, 1996)
38. B. Edde, *RADAR Principles, Technology, Applications* (Englewood Cliffs, NJ, Prentice Hall, 1995)

39. E.A. Theodorou, M.R. Gorman, P.R. Rigg, F.N. Kong, Broadband pulse-optimized antenna, IEE Proc. H **128**(3), 124–130 (1981)
40. S. Evans, F.N. Kong, TEM horn antenna: input reflection characteristics in transmission. IEE Proc. H **130**(6), 403–409 (1983)
41. J.D. Cermignani, R.G. Madonna, P.J. Scheno, J. Anderson, Measurement of the performance of a cavity backed exponentially flared TEM horn, in *Proceedings of SPIE, Ultrawideband Radar,* vol. 1631 (1992), pp. 146–154
42. C. Nguyen, J.S. Lee, J.S. Park, Novel ultra-wideband microstrip quasi-horn antenna. Electron. Lett. **37**(12), 731–732 (2001)
43. J.S. Park, C. Nguyen, Low-cost wideband millimeter-wave antennas with seamless connection to printed circuits, in *2003 Asia Pacific Microwave Conference* (Seoul, Korea, 2003)
44. J. Han, C. Nguyen, Investigation of time-domain response of microstrip quasi horn antennas for UWB applications, IEE Electron. Lett. **43**(1), 9–10 (2007)
45. Ansoft High-Frequency Structure Simulator (HFSS), *Ansoft Corporation*, Mequon, Wisconsin
46. F.E. Churchill, G.W. Ogar, B.J. Thompson, The correction of I and Q errors in a coherent processor. IEEE Trans. Aerosp. Electron. Syst. **17**(1), 131–137 (1981)

Index

C. Nguyen and J. Park, *Stepped-Frequency Radar Sensors*,
SpringerBriefs in Electrical and Computer Engineering,
DOI 10.1007/978-3-319-12271-7

Printed in the United States
By Bookmasters